Fundamentals of Marine Riser Mechanics

Fundamentals of Marine Riser Mechanics:

Basic Principles and Simplified Analyses

Charles P. Sparks

DISCLAIMER

The contents of this book and the Excel files on the accompanying CD-ROM are intended to allow the reader to explore and better understand marine riser behavior. The author and the publisher will assume no liability whatsoever for any loss or damage that may result from the use of any of the material in this book or of any of the Excel files. Such use is solely at the risk of the reader.

PennWell Corporation
1421 South Sheridan Road
Tulsa, Oklahoma 74112-6600 USA

800.752.9764
+1.918.831.9421
sales@pennwell.com
www.pennwellbooks.com
www.pennwell.com

Director: Mary McGee
Managing Editor: Marla Patterson
Production Manager: Sheila Brock
Production Editor: Tony Quinn
Cover Designer: Clark Bell
Book Designer: Brigitte Pumford-Coffman

Library of Congress Cataloging-in-Publication Data

Sparks, Charles P.
Fundamentals of marine riser mechanics : basic principles and simplified analyses / Charles P. Sparks.
-- 1st ed.
p. cm.
ISBN 978-1-59370-070-6
1. Offshore oil well drilling. 2. Drilling platforms. 3. Oil fields--Production methods. I. Title.
TN871.S653 2007
622'.33819--dc22

2007023688

Printed in the United States of America

1 2 3 4 5 11 10 09 08 07

To Brigitte, Marie-Anne, and Xavier

Contents

Preface

This book is the result of 30 years of fascination with riser behavior. It all began in 1976, when, as a young engineer with a background in civil engineering, I first attempted to design a production riser for a floating platform, with the help of a rudimentary computer program. Initial attempts led to a riser that was grossly overstressed in bending. To my astonishment, increasing the bending stiffness of the overstressed sections only increased the moments and left the bending stresses virtually unchanged. The riser was clearly a strange structure that did not react at all in accordance with the standard intuition of a young civil engineer. It was only with the help of simplified analytical calculations that I began to understand why the riser behaved in such a strange way.

The next topic of fascination was the influence of pressure on pipe and riser stability and deflections. I quickly discovered that the subject was very tricky. Nevertheless, I was impressed and almost reassured to note that virtually everyone else working in the offshore industry found the subject to be tricky too. It even provoked passion! Although there were plainly several different ways of approaching the subject, all of which led to the same equations and conclusions, some engineers defended their particular approach with great vigor, as though people who took a different line were guilty of thought crime! Such passion was all the more surprising since the subject is, in fact, elementary, in the sense that its study requires no advanced mathematics. It was also fascinating that very similar problems occur in civil engineering (in the analysis of concrete beams and columns) that lead to exactly the same equations. Yet, among civil engineers, the subject provokes neither passion nor any discussion even.

Many other areas of fascination followed. To name a few, these included the stability of drilling-riser kill and choke lines; axial resonance of hung-off risers; design of top-tensioned risers for tension leg platforms and spars; the potential of new materials such as high-performance composites for improved riser design; multiple barrier risers; stress joints; riser bundles; steel catenary risers. Many of these are treated in this book.

During my career, I had neither the time nor the means to delve into these subjects as deeply as I wished. Work priorities interfered with all that! More recently, retirement has provided the time, and Excel files have provided the means.

Analytical techniques, which were so helpful in understanding the influence of bending stiffness on riser behavior, have been used copiously throughout this book. The reader should not, however, conclude that I advocate the abandonment of numerical analysis. Quite the contrary. The ability of today's numerical computer programs to take into account a vast number of parameters and give immediate results of high precision is already something to marvel at. In the future, such programs will doubtless become increasingly sophisticated. Someday they may be able to take into account the sunspot cycle, the effects of global warming, and maybe even the humor of the riser operator! There is no reason why such programs should not be used. However, numerical analysis is not good at explaining why particular results are obtained. The more sophisticated the calculation used, the less obvious it is which parameters are really significant and how those parameters influence results.

Analytical methods may require some slight simplification of the problem to be applicable, but they generally lead to very compact formulae that do explain which, why, and how parameters influence results. Furthermore, those compact formulae are often very simple to program and hence can be useful for preliminary analyses. Those two reasons are the principal interest and justification for applying analytical methods.

To cite just one example, it is shown in chapter 12 that if two catenaries with and without bending stiffness EI (that are otherwise identical) are compared, the difference in the position of the touchdown points (TDPs) with respect to the top end is equal to the length $\sqrt{EI/H}$, where H is the horizontal component of the effective tension. Further analysis shows that the shear force at the TDP of the stiff catenary is equal to the apparent weight of that length. Thus, the top tension is (virtually) the same even though there may be a significant difference between the suspended lengths. Analytical expressions often lead directly to this kind of insight, which is much more difficult to obtain from numerical results. Similar ultrasimple analytical expressions and deductions occur in many of the chapters of this book.

Analytical methods can also be used in some cases to show that sophisticated numerical analyses of seemingly very complicated problems may not even be necessary. For example, it is shown in chapter 10 that the distribution of the riser global moments between the different tubes in a riser bundle can be expressed analytically very simply, even though that distribution is absolutely not in simple proportion to the bending stiffnesses of the individual tubes. Consequently, the moments in the different tubes can be obtained directly from the global analysis. The alternative is to model every tube in the bundle and hope that the program is capable of analyzing correctly the vast number of interactions between them.

Excel files have been used to verify the findings included in this book, as well as to generate figures and tables. It has not been difficult to adapt the files for more general use, and 17 Excel files are included on the accompanying CD-ROM. For most of them, the results of different independent calculation methods are compared. For example, some files compare results of analytical calculations, made using some simplifications, with those made using an "exact" numerical method—the object being to demonstrate the precision of the analytical method and justify the conclusions drawn from it. The Excel files allow the reader to test those conclusions with different data.

The pleasure it has given me to research and write this book is immense. I hope that the early chapters, on the effects of pressure, will be helpful to young engineers approaching the subject for the first time and to those who still have difficulty with it. I also hope that the same chapters may be useful to those who have no difficulty with the subject but who are frequently called on to explain it to others less fortunate than themselves. Further, I hope that the rest of the book will help all readers to better understand riser static and dynamic behavior and allow them to appreciate which, why, and how parameters influence the results they do. Above all, I hope that the book will be an enjoyable read.

Charles Sparks,

February 2007

Acknowledgments

First, I wish to thank the directors of the Institut Français du Pétrole for allowing me and encouraging me to spend such a large part of my career exploring so many different aspects of riser behavior. I particularly wish to remember Jacques Delacour and to thank Jean-François Giannesini, Jacques Burger, Jacqueline Lecourtier, and Christian Pauchon.

Many colleagues at the Institut Français du Pétrole helped in different ways with the contents of this book, by carrying out simulations, by providing comments on sections, and by discussing the best ways of presenting particular topics. I particularly wish to thank Jean-Marie Alliot, Daniel Averbuch, Francis Biolley, Cedric Le Cunff, Jean Falcimaigne, Emmanuel Fontaine, Philippe Gilbert, Jean Guesnon, and Pierre Odru.

I also wish to thank Irv Brooks, Bill Hudson, Alain Marion, Howard Shatto, Charles White, and Jerry Williams for their valuable help and comments on different chapters.

Most special thanks are due to Chris Mungall for his confirmatory simulations, for his copious comments on many chapters, and for his ideas and suggestions for modifying and enlarging the text to make it easy to follow for readers who have come to riser engineering from other backgrounds than my own.

I also wish to acknowledge help from friends outside the offshore industry. Barry Clegg and George Kistruck contributed valuable comments on the style of the text. The former also suggested ways of improving the presentation of the Excel files, based on his expertise in technical communication. Iain Kennedy has helped with some of the mathematics.

Finally, I wish to express my gratitude to Xavier Roulhac de Rochebrune for solving the equations of Appendix I.

Nomenclature

The following list presents the principal nomenclature. The number of letters being limited, some symbols have different meanings in different chapters. The precise meaning of each symbol is always explained in detail in the text. Likewise, numerous subscripts are used and defined in the text.

a: amplitude

A: cross-sectional area

b: SJ constant

B: constant

c: celerity

C_D: drag coefficient

C_m: added mass coefficient

C_M: inertia coefficient

d: depth

e: stretch; element length

E: Young's modulus

EA: axial stiffness

EI: bending stiffness

f: load function; flow-line length

F: force

F_w: apparent weight function

g: gravitational acceleration

G_{pt}: pressure-temperature function

G: beam curvature function

H: beam function;
horizontal force

I: second moment of area

k: stiffness; $\sqrt{T / EI}$

K: damping parameter

L: length

m: mass per unit length

M: moment; mass

n: mode number

p: pressure

P: axial force

P_E: Euler buckling load

q: w/kT

Q: equivalent tension

r: radius

R: radius of curvature; damping parameter

s: catenary suspended length

S: total length; damping parameter

t: temperature; time

T: tension

T_p: vibration period

u: velocity; displacement

U: upthrust; vertical amplitude

v: velocity

V: vertical force

w: weight per unit length

W: weight

x: coordinate

y: coordinate

Y: lateral amplitude

z: height; setdown

α: thermal expansion coefficient;
SJ constant; damping parameter

β: bundle moment parameter;
damping parameter

γ: riser angular offset; λ' / mc

ε: strain

θ: angle

ψ: angle

λ: damping coefficient

ν: Poisson's ratio

ρ: mass density

σ: direct stress

τ: shear stress

ϕ: diameter; damping parameter

ω: circular frequency ($2\pi/T_p$)

SI Unit Equivalents

The examples given in the text and the accompanying Excel files are all presented in homogenious SI units, using meters, kilonewtons, kilopascals, and tonnes. By contrast, pipe diameters and wall thicknesses are input in inches, in accordance with the usual practice.

For those who are less familiar with SI units, the following equivalents may be helpful. Except as otherwise noted, all equivalents are rounded values.

• Length:	1,000 ft	= 304.8 m (exact);
	10 in.	= 0.254 m (exact)
• Force:	1 kip	= 4.45 kN;
. Force/length:	1 kip/ft	= 14.59 kN/m
• Pressure/stress:	1 ksi	= 6,895 kPa (Note: 1 kPa = 1 kN/m^2)
• Bending stiffness:	1 kip-in.^{2}	= 0.00278 kN-m^2
• Moment:	1 kip-ft	= 1.36 kN-m
• Mass:	1,000 lbs	= 0.4536 tonnes (Note: 1 tonne = 1,000 kg)
• Mass/length:	1,000 lbs/ft	= 1.488 tonnes/m
• Density (mass):	1 lb/ft^3	= 0.0160 tonnes/m^3
• Velocity:	1 knot	= 0.514 m/s

The following are useful SI values:

- Young's modulus of steel: = 210×10^6 kPa
- Young's modulus of titanium: = 110×10^6 kPa
- Density (mass) of steel = 7.84 tonnes/m^3
- Density (mass) of seawater = 1.025 tonnes/m^3
- Steel pipe with 21 in. outer diameter and 1 in. wall:

  Axial stiffness: EA = 8,512,700 kN

  Bending stiffness: EI = 275,290 kN-m^2
- Gravitational acceleration, g = 9.81 m/s^2

1 Introduction

This book is principally aimed at explaining the way marine risers behave. It begins with a brief review of the different types of risers that are in use today, with some history and illustrations, as well as references to the types of vessels with which they are associated. Then, an overview of the contents of the following chapters and appendices will be given.

Riser Types

Marine risers date from the 1950s, when they were first used to drill offshore California from barges. An important landmark occurred in 1961, when drilling took place from the dynamically positioned barge CUSS-1. Since those early days, risers have been used for four main purposes:

- Drilling
- Completion/workover
- Production/injection
- Export

Within each group, there is immense variety in the detail, dimensions, and materials, as explained in the following subsections. Drilling risers can be subdivided into low-pressure and high-pressure risers.

Production risers, used from floating platforms, inevitably followed some years after drilling risers. They were first used in the 1970s with an architecture inspired by that of top-tensioned drilling risers. Since then, they have taken many other forms, including bundled risers, flexible risers, top-tensioned risers (TTRs), steel catenary risers (SCRs), and hybrid risers, which are a combination of steel and flexible risers.

Export risers also come in a variety of architectures similar to those of production/injection risers, but generally with larger diameters and lower pressures. Hence, in the following subsections, production/injection risers and export risers are categorized by their architectures.

Low-pressure drilling risers

The standard drilling riser today is a low-pressure riser, open to atmospheric pressure at the top end. Thus, the internal pressure can never exceed that owing to the drilling-mud weight. Drilling risers are made up of a number of *riser joints,* typically 15–23 m long, as illustrated in figure 1–1, which shows a riser joint being prepared for running on a drill ship.[1] A typical drilling riser comprises a central tube of 21 in. nominal diameter and is equipped with a number of peripheral lines. Four such lines are shown in figure 1–1: kill and choke lines, used to communicate with the well and circulate fluid in the event of a *gas kick* for which the seafloor blowout preventer (BOP) has to be closed; a booster line, used to injected fluid at the low end of the riser and accelerate the flow so as to better evacuate the cuttings; and a small-diameter hydraulic line, used to power the seabed BOP.

The joint shown in figure 1–1 is also equipped with syntactic foam buoyancy modules, to reduce the weight in water. Drilling risers are generally equipped with such modules over the upper part of their length. A short length close to the surface is usually left bare, to reduce hydrodynamic loads in the zone where the wave forces are greatest. The lower part of the riser is also often left bare, since the density (and cost) of syntactic foam increases with the required design pressure—and hence with depth.

In the past, air-can buoyancy units have sometimes been used. These had the advantage of allowing the buoyancy to be adjusted and optimized for each drilling campaign, but they introduced an additional level of complexity.

A further feature of the riser joint is the connector, which can be of different types. Figure 1–1 shows a *breechblock connector*. Figure 1–2 shows a further example of a drilling-riser joint, on which some of the peripheral lines can be seen more clearly. The joint has a flanged connector and is unusual in that it is made of aluminum.[2]

Fig. 1–1. *CLIP drilling-riser joint (courtesy of the Institut Français du Pétrole)*

Fig. 1–2. *Aluminum drilling-riser joint (courtesy of Aquatic, Moscow)*

As shown in figure 1–2, the peripheral lines are attached to the main tube by several guides. These have to be carefully designed, since they prevent the peripheral lines from buckling under the effect of internal pressure. As a backup safety system, it is also good practice to design the lines so that they cannot break out of their housing at the connector level, even if the line should buckle.

Figure 1–3*a* and *b* show drilling risers deployed below a drill ship and a semisubmersible, respectively, which are generally called mobile offshore drilling units (MODUs).

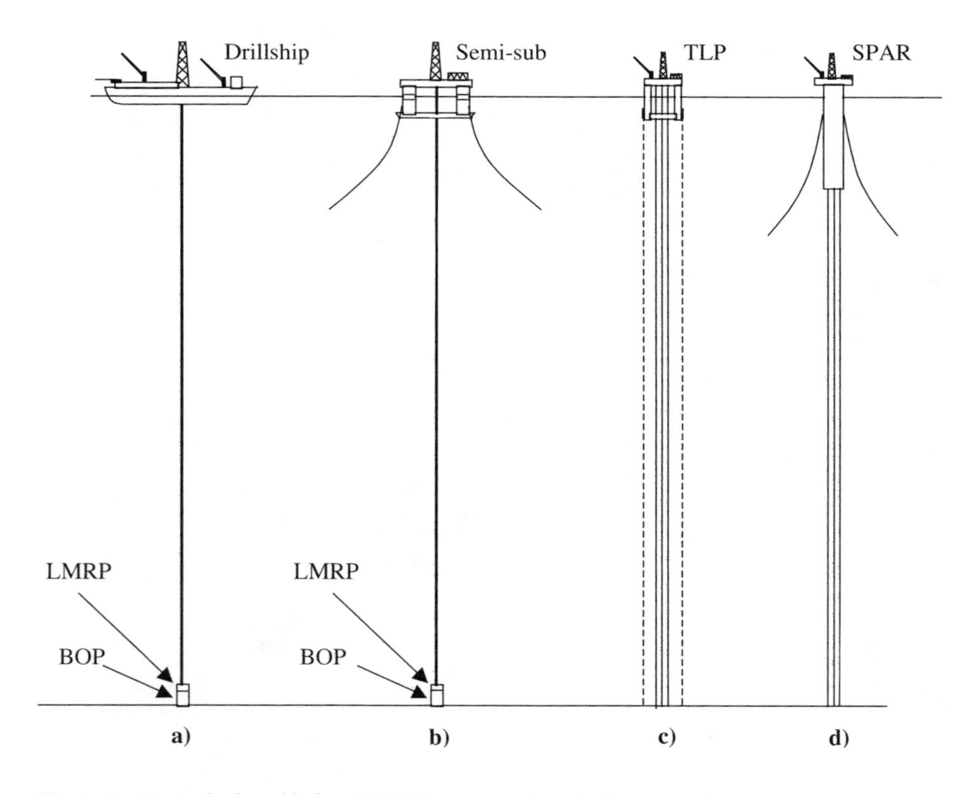

Fig. 1–3. *Risers deployed below MODUs, tension leg platforms, and spars*

The seabed BOP allows the drill string to be cut and the well to be closed in case of an emergency, and the lower marine riser package (LMRP) then allows the riser to be disconnected. A flex joint at the junction between the riser and the LMRP allows limited rotation of the riser and avoids concentrated moments. Weights of the BOP and LMRP depend on the design pressure, but are on the order of 100–300 tonnes for the BOP and 50–100 tonnes for the LMRP.

High-pressure drilling risers

When the BOP is located at the surface, a high-pressure drilling riser is required, as was the case for the CUSS-1 in 1961. This riser has a much simpler architecture than does a low-pressure riser, since it does not require kill and choke lines. In the event of a gas kick, the BOP is closed at the surface, which is immediately accessible on the drilling rig. Thus, the riser has to be designed to take the full well pressure. However, there is potentially more risk when drilling with a surface BOP, unless an adequate seabed disconnection system can be provided in case of an emergency.

Following the CUSS-1, the Sedco 135A continued to drill with a high-pressure riser in shallow water (<75 m) for many years. High-pressure risers with surface BOPs have been used again since the 1980s to drill from many tension leg platforms (TLPs), such as Hutton (1984), Heidrun, Mars, RamPowell, and URSA, and from some spars. In the case of the Heidrun TLP, the drilling riser was made of titanium.

A high-pressure slimline (small-diameter) riser with surface BOP was also proposed for the Ocean Drilling Program in the early 1990s, to allow scientific drilling with mud circulation in ultradeep water (>4,000 m).[3] However, the project was not pursued. More recently, high-pressure risers with surface BOPs have been used to drill a large number of wells from semisubmersibles in moderate environmental conditions.[4] The concept continues to be developed for deeper water and harsher environments.[5]

Completion/workover risers

Completion/workover risers have similarities with both high- and low-pressure drilling risers. They come in a variety of diameters, intermediate between that of a simple drill string and a full-scale drilling riser, depending on the work that has to be performed. They feature a high-pressure design with the wellhead on the platform and therefore do not require kill and choke lines. When such a riser is used from a MODU, the lower end is designed to allow an emergency disconnect. A riser safety package (RSP) on the seabed allows the well to be closed, and an emergency disconnect package (EDP) allows the disconnection. Power to those packages is provided by an umbilical attached to the riser.

Bundled risers

The term *bundled riser* encompasses several different riser architectures. The first production riser installed on a floating platform was used on the Argyll field, in the North Sea, in 1975. It was of the bundled type and was inspired directly by the architecture of low-pressure drilling risers. It comprised a core pipe with a number of satellite production risers around it, to which it was attached by guides. These guides were equipped with funnels, so that the satellite risers could be run down individually from the platform, once the core pipe was in place. The central core pipe also served as the *export* riser. The system worked well but was complicated since each satellite riser had its own tensioner.

A bundled riser was used again in the late 1980s on the Placid field. This riser consisted of a core pipe that served as guide for about 50 satellite lines. Instead of using tensioners, the Placid riser was tensioned by a combination of syntactic foam buoyancy modules and a subsurface buoy, situated immediately below the semisubmersible production platform. The core pipe terminated at the lower end in a *stress joint* made of titanium, to give it increased flexibility. Since the satellite risers were freestanding (in compression), they had to be run through guide tubes at the stress joint level to prevent buckling. At the top end, they were connected to the platform by flexible flow-line jumpers.[6]

The same concept has been used more recently in the form of the *hybrid riser towers* (fig. 1–4) on the Girassol and Greater Plutonio developments in West Africa.

These riser towers are used in association with floating production, storage, and off-loading platforms (FPSOs). They are no longer installed below the platform as on Placid, but are installed to one side of the FPSO, to which they are connected by flexible lines. The satellite risers are suspended from the buoyancy tank.

This architecture results in a curious situation near the top end. The satellite risers are attached to the core pipe by guides, which maintain them in position. However, once the satellite risers are in place, suspended from the buoyancy tank, their weight causes the core pipe to go into effective compression over a considerable length below the tank. Buckling is prevented by the tension in the satellite risers. This sounds like a dangerous conundrum in which the satellite risers and the core pipe each rely on the other for support. However, the riser is stable since the global bundle *effective tension* is positive.

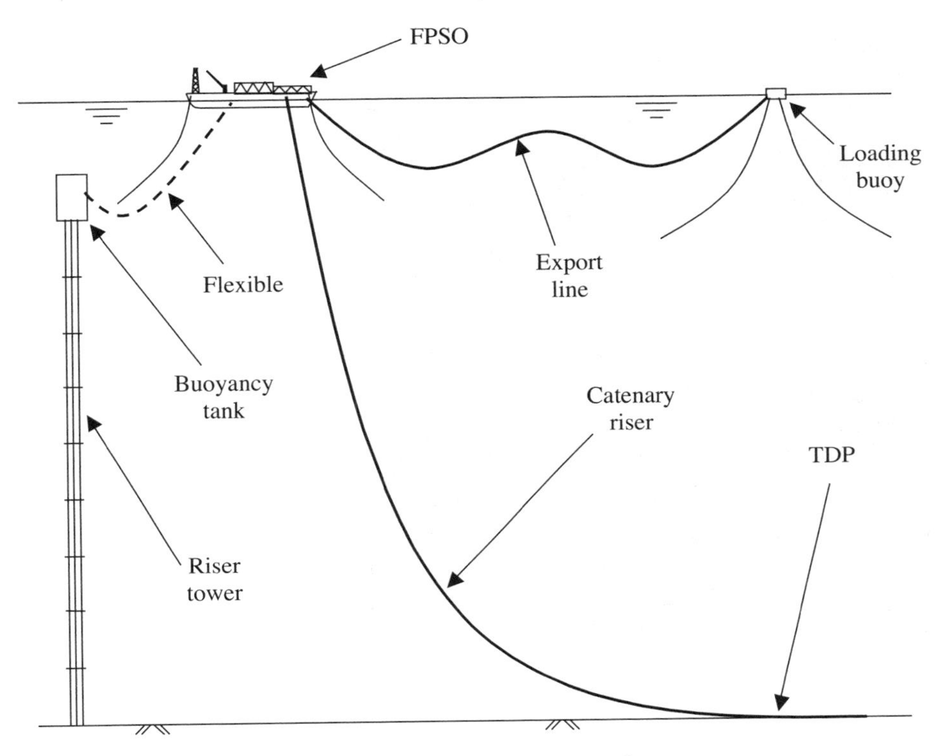

Fig. 1–4. *Riser towers, catenary risers, and mid-depth export lines*

Flexible risers

Flexible risers are the result of an extraordinary development program. During the 1960s, the Institut Français du Pétrole attempted to develop a flexible drilling system with a down-hole turbo-drill, termed *Flexoforage*. The development continued throughout the 1960s, with much of the testing being done in cooperation with Russian engineers at Otradny, near Kouïbychev, on the Volga. The Flexoforage application finally failed because it was not possible to develop a safe system (equivalent to the BOP) for closing the well in the event of a blowout. However, flexible pipes had by then been developed, and other applications were immediately sought. Flexible pipes were found to be ideally suited to offshore applications in the form of production and export risers, as well as flow lines. These applications were fully exploited, from the 1970s onward, by Coflexip. Flexible pipes are characterized by their wall structure, as shown in figure 1–5.

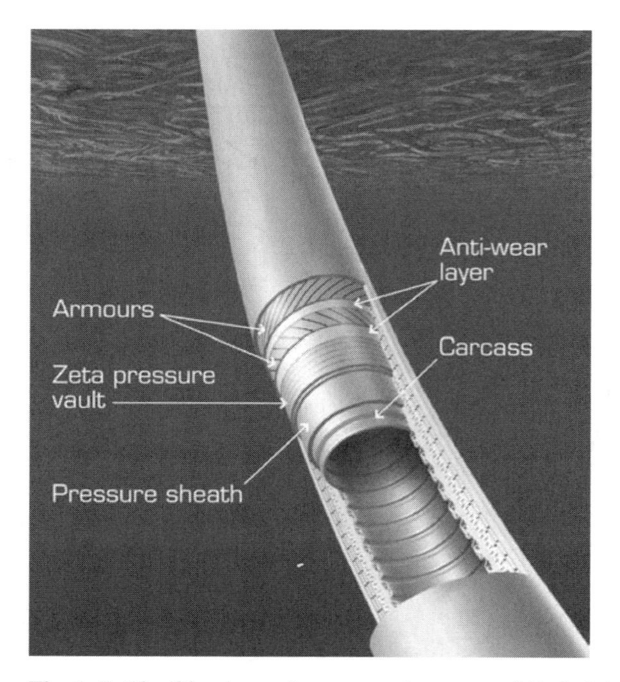

Fig. 1–5. *Flexible-pipe wall structure (courtesy of Technip)*

In the case of a flexible pipe, the many different functions of a standard metal pipe wall are taken by different layers, as shown in figure 1–5. Typically, the structure includes an inner metal carcass for collapse resistance; a plastic pressure sheath for fluid containment; a steel vault for hoop stress resistance; steel armors to resist axial tensile loads; and a plastic outer sheath to prevent seawater penetration. The resulting structures are complex, and their study is beyond the scope of this book.

Flexibles pipes were first used as flow lines in the early 1970s and then as dynamic risers, in 1977, on the Garoupa field, offshore Brazil. Applications to other Brazilian fields followed. In the early 1980s, they were used as dynamic risers in the North Sea, on the Duncan and Balmoral fields.

Model tests showed that flexible pipes could be used in the *free-hanging* mode, but only in benign environments. In more severe environments, changes in lateral offset of the platform tended to induce unacceptably large variations in pipe tension, and large changes in position of the touchdown point (TDP) at the seabed. To reduce those effects, a number of different architectures were developed for the section close to the seabed. Some of these are shown in figure 1–6.

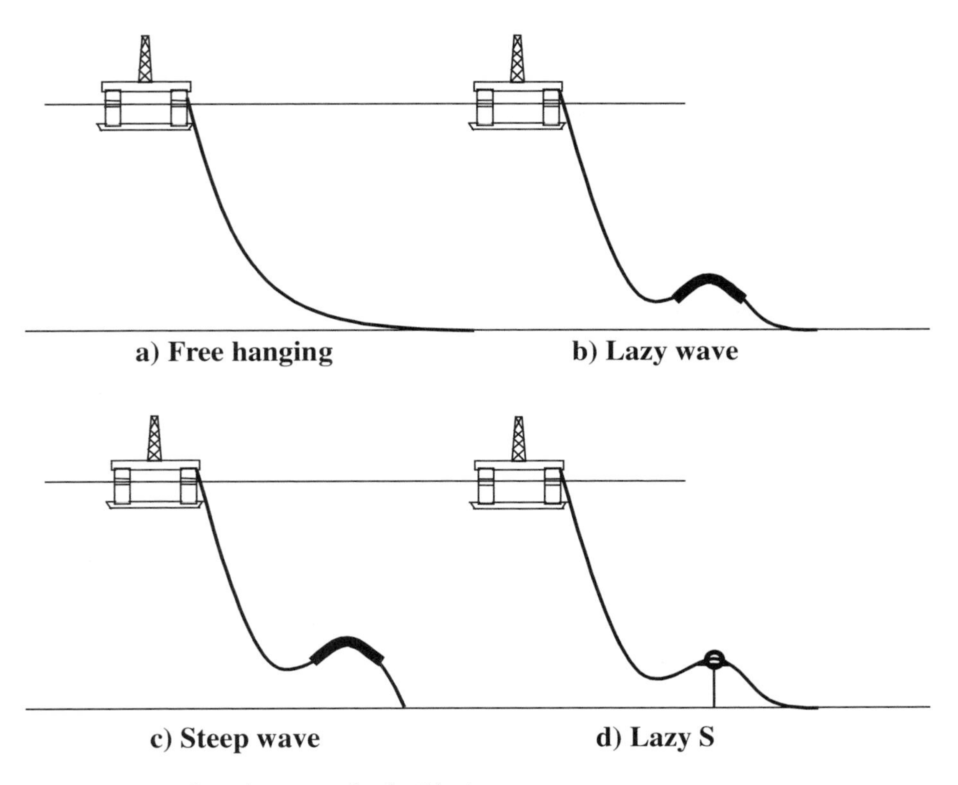

Fig. 1–6. *Example architectures for flexible risers*

Today flexible pipes have been used as risers and flow lines in water depths down to 2,000 m.[7] Offshore Brazil in the Campos Basin alone, approximately 4,000 km of flexible pipes have been deployed.[8]

Individual top tensioned risers (TTRs)

Individual risers that rely on a top tension in excess of their apparent (wet) weight for stability are generally termed TTRs, although all risers require top tension. TTRs were first used in 1984 for production/injection and export on the Hutton TLP.[9] They have since been used on many other TLPs and spars. The spacing between such risers is typically small, as shown in figure 1–3*c* and *d*. TTR production/injection risers are generally designed to give direct access to the well, with the wellhead on the platform; the riser has to be capable of resisting the tubing pressure, in case of a tubing leak or failure. Connection to the seabed is by a stress joint, which is sometimes made of titanium for increased flexibility. In the case of spars, each riser is also equipped with a *keel joint,* designed to relieve bending stresses at the spar keel level.

In general, TLP risers use tensioners, whereas spar risers are tensioned by air cans, athough one spar (Holstein) uses tensioners. Top-tension ratios (ratio of top tension to riser apparent weight) vary greatly. On some TLPs, they have been designed to be as low as 1.2, while on others they have been in the range of 1.4–1.8. There is one significant difference between tensioners and air cans, resulting from the setdown of the riser top end as the platform is offset. In the case of tensioners, the top tension *increases* with platform offset because of tensioner stiffness. In the case of air cans, it is slightly *reduced* because of the compressibility of the air or gas in the cans.

For both TLPs and spars, the spacing between riser centers is on the order of 3–6 m, without any guides between them. Possible interaction between adjacent risers because of slightly different profiles has to be studied carefully. Such interaction can result from the combined effects of platform offset, current load, differences in top tension and apparent weight, thermal effects, and vibrations. It is particularly critical when adjacent risers have quite different characteristics and functions (production and drilling).

Some TLPs and spars are equipped with single-barrier risers (plus production tubings). Examples are Hutton, Jolliet, Heidrun, and Marlin, whereas others, such as Auger, Mars, RamPowell, Ursa, and Brutus, are equipped with dual-barrier risers. The latter are inevitably of larger diameter, which aggravates the interaction problem. They are also very heavy, particularly since the annular area between the two riser casings is generally filled with drilling mud, designed to kill the well in case of failure of the inner riser casing.

Steel catenary risers (SCRs)

SCRs are a recent addition to the riser family (e.g., see fig. 1–4). SCRs were initially used as export lines on *fixed* platforms. Their first application to a floating platform was as export risers on the Auger TLP (1994).[10] Since then, they have been used in progressively more severe applications. SCRs were used in the late 1990s, again for export, on the P-18 semisubmersible offshore Brazil. More recently, they have been used in large numbers as production/injection risers associated with the Bonga FPSO.

SCRs plainly have similarities with free-hanging flexible risers and inevitably suffer from the same problems: large tension fluctuations and large movements of the touch down point (TDP), leading to fatigue, induced by platform motions and changes in offset. These problems cannot be solved as easily as they were for flexible risers, but the possibility of adding buoyancy near the TDP to give the SCR something approaching a lazy-wave–type architecture is under study. Likewise, the interaction between the riser and the soil at the TDP is the subject of research.

Mid-depth export lines

Figure 1–4 also shows a sketch of a mid-depth export line—which is not a riser, even though there are similarities. Such export lines are used to transport the production from an FPSO to a loading buoy. They are typically about 2 km long, on the order of 16–22 in. diameter, and equipped with buoyancy modules over a central length that maintains them at *mid-depth* with a W shape. The shape and depth of the line is influenced by the density of the internal fluid. Such export lines have been made from steel pipes (Girassol, Kizomba, Erha) and from flexible pipes (Bonga).

Overview

The remaining content of this book is divided into four chapters on the effects of pressures, seven chapters on riser statics, three chapters on riser vibrations, and 10 assorted appendices. While the chapters on the effects of pressures (chaps. 2–5) are based on an earlier publication,[11] they have been expanded to take into account 20 years of comments, as well as further thoughts and discussion about the matter. Although the subject is "elementary," those chapters are probably the most important of the entire book, since there is still the potential for catastrophic misunderstandings about the effects of pressure. Different ways of deriving the effective tension equation are presented, and numerous difficulties provoked by the subject are discussed in detail. A procedure for decomposing the effective tension (and the apparent weight) in a multi-tube riser (consisting of separate tubes and/or tubes within tubes) into components from the individual tubes is explained, to allow straightforward analysis of very complicated problems involving changes of internal fluids, temperatures, and pressures.

The chapter on wall stresses in a circular cylindrical elastic pipe (chap. 4) defines the concept of effective stress, which is numerically equal to the pipe effective tension divided by the wall section. It is shown that such a stress is equal to the excess axial stress in the pipe wall (i.e., in excess of the axial stress that would be induced in the pipe by internal and external pressures, if it were closed either side of the section concerned). It is shown that the concept of effective stress leads to the simplification of many stress problems.

Ways of calculating axial strains for both isotropic and anisotropic pipes are given in chapter 5. The latter are defined in terms of equivalent Poisson's ratios, which can be easily determined from laboratory tests on tube samples. Axial stretch due to changes of internal fluid and pipe upending (following tow-out) is then explained, as are ways to obtain the axial stretch of multi-tube risers owing to internal changes of temperature, pressure, and fluid density. The calculation procedure requires the effective tension to be decomposed into its component parts from each tube, as explained in chapter 3.

Chapter 6 is pivotal since its results are exploited in most of the chapters that follow. It examines the behavior of beams under constant tension, for which the basic deflection equation can be solved analytically. A riser with zero apparent weight is a beam under constant tension. This chapter allows important conclusions to be drawn about the influence of bending stiffness on curvature, end angles, end rotational stiffness, and end shear. Chapter 7 then formulates the simple equations of tensioned cables, with weight but zero bending stiffness.

In chapter 8, the conclusions of chapter 6 are tested for risers with both bending stiffness and apparent weight. Excel files are used to compare results given by the cable equations of chapter 7, with "exact" values obtained numerically. It is shown that the conclusions of chapter 6 are valid, with some minor adjustments, for risers with both bending stiffness and apparent weight, even though those conclusions cannot be proved analytically.

Chapter 9 applies some of the results of chapter 6 to the design of stress joints. It shows how the evolution of the bending stiffness, along the joint, can be chosen to give a particular bending effect such as circular bending, or constant bending stresses on the joint outer surface. An Excel file allows the resulting behavior of the joint to be checked numerically.

Chapter 10 looks at the local bending behavior of individual pipes, between guides, in a riser bundle. It is shown that the total moment in the bundle is given correctly by global analysis in which the bundle is simulated as a single structure with the combined characteristics of all the tubes in the bundle (total effective tension, total apparent weight, and total bending stiffness) subject to the sum of lateral loads on all the pipes. However, the distribution of the bending moments between the different pipes depends on several factors, including the load type (apparent weight, hydrodynamic or inertia forces) and the effective tension distribution between the pipes. A further Excel file gives the distribution of moments according to load type, for different data.

Chapter 11 is devoted to TTRs and shows how riser tension and sag evolve with platform offset, particularly when the risers are associated with TLPs or floating platforms. It is shown how riser behavior is influenced by tensioner stiffness, as well as by internal changes to the riser temperature, pressure, and fluid densities. Importantly, even internal changes in a tubing can influence the riser behavior, unless the tubing is equipped with a special, balanced expansion joint designed to avoid such effects.

Chapter 12 looks at the behavior of SCRs. The results obtained in chapter 6 are applied to parts of the catenary. It is deduced that bending stiffness has negligible effect on top tension and on the horizontal component of effective tension (H), for a given total horizontal projection of the total length (suspended part plus seafloor flow line), although the position of the TDP is changed. A simple expression for the change in position of the TDP is given in terms of the bending stiffness and the horizontal force (H). It is also shown that the curvature of a stiff catenary is *greater* than that of a pure catenary over most of its length, which may initially be surprising. Results of numerical simulations are included, which confirm those predictions.

Chapters 13–15 are devoted to riser vibrations. Axial and transverse vibrations are both the result of stress waves that continually ascend and descend a riser. Resonant periods are always equal to twice the time for a stress wave to run between adjacent nodes. Likewise, the time for stress waves to run between adjacent nodes and antinodes are all equal, even when the stress waves are not propagated at constant velocity.

Chapter 13 looks at axial vibration of risers fixed to the seabed at the lower end. Such vibrations are of little consequence in the real world, but they have the merit of being simple to study, thereby serving as introduction to the two chapters that follow.

Axial vibrations of hung-off risers are studied in chapter 14. These are particularly important for drilling risers in the disconnected mode. Simple formulae for their natural periods are given. It is also shown that, if damping is not extremely large, the response at resonance depends only on the quantity of energy removed per cycle. It makes little difference whether the damping is distributed over the length of the riser or even concentrated at an equivalent point well below the riser bottom end.

Chapter 15 is devoted to transverse modal vibrations of the type induced by lock-in of vortex-induced vibrations (VIV) in near-vertical risers. Five different ways of analyzing the problem, based on different degrees of approximation, are discussed. Simplified analytical methods lead to a number of very simple formulae for many characteristics including the natural periods, node positions, maximum curvature, and bottom-end angle. An Excel file allows the results of four simplified methods to be compared with the 'exact' results given by numerical simulations.

The book also includes 10 appendices. Appendix A is devoted to deriving the standard fourth-order differential equation that applies to a tensioned beam, such as a riser. The equation can seem mysterious without explanation, but becomes straightforward when derived from first principles.

Appendix B is a supplement to chapter 2. Alternative ways of deriving the effective tension equation for certain simple riser cases are shown.

Morison's equation as applied to moving bodies in accelerated flow, such as risers subject to wave and current action, is the subject of appendix C. This equation also can seem mysterious without explanation, but again becomes very clear when derived from first principles.

Appendix D derives the standard Lamé equations for the pressure-induced stresses in a thick-walled circular cylindrical elastic pipe. They can be found in references and are included here for the convenience of the reader.

Appendix E shows how the equivalent Poisson's ratios defined in chapter 5 for anisotropic pipes are related to the basic properties of the anisotropic pipe wall material. It is also shown how the equivalent Poisson's ratio for the external pressure can be deduced from the axial and external circumferential strains measured in an axial tension test.

Appendix F derives the basic equation of a tensioned beam subject to generalized load. It is a supplement to chapter 6.

Appendix G is devoted to the moments in individual pipes in a riser bundle. The equation for the moments in a pipe between guides is related to the global moment in the bundle. It is derived for individual pipes in effective tension and in effective compression. It is also given for the limiting case of a pipe with zero effective tension.

Appendix H derives the standard catenary equations (with zero bending stiffness), which can be found in many references. They are included for the convenience of the reader. Appendix H also gives the consequences of axial stretch on the catenary top-end coordinates. A numerical formulation for the standard large-angle differential equation for catenaries with bending stiffness is also given.

Appendix I gives the solutions to the response equations of a hung-off riser with damping. Two cases are given—one with distributed damping along the riser plus concentrated damping at the lower end and another for damping concentrated at an equivalent point below the riser lower end (which is much simpler to model). Appendix I also shows how the magnitude of the damping at the equivalent point can be chosen to give the same mean rate of energy dissipation per cycle as for distributed damping.

Appendix J is devoted to explaining points about the Excel files provided on the CD-ROM that accompanies this book. It also explains the solution method employed in the numerical simulations in many of the files.

Excel File *SCR-Example.xls*

This book derives and presents a great many equations and expressions that can be applied by the reader to the preliminary analyses of many particular riser problems. They are all simple to program. As an example, this section shows how those equations and expressions have been applied to the analysis of SCRs in the Excel file *SCR-Example.xls,* provided on the CD-ROM. The file is representative of calculations that are typically made by a riser designer at the start of a project. The file provides an overview of the input to the problem and a number of significant results. The reader can use the file to explore the behavior of an SCR using different data. The reader may find it useful to return to the file when advancing through the book, particularly when reading chapters 4, 12, and 15. To allow easy printing, a further version of the file, named *SCR-Example (print).xls,* is also provided.

For preliminary SCR design, a designer will need to calculate the riser catenaries with the top end in the "mean," "near," and "far" offset positions, in order to check the tensions, positions of the TDPs, and top-end angles. The designer will also want to check the various maximum stresses in the upper regions and near the TDP and to compare results with values permitted by relevant design codes. For preliminary analyses, he or she is likely to have sufficient data to give the following:

- Pipe diameter and wall thickness
- Pipe material modulus and yield stress
- Water depth
- SCR total length (including seafloor flow-line length)
- Densities (internal fluid, pipe material, external [seawater] fluid)
- Top-end internal pressure
- Top tension

From the densities and the pipe dimensions, *SCR-Example.xls* calculates the pipe apparent weight. For a real design, this may need to be modified to take into account possible coatings, insulation, and buoyancy modules.

Using as data only the top tension and apparent weight (w), the file then calculates values for the *cable* catenary. (Note that the word "catenary" is used loosely in the offshore industry. To avoid confusion, catenaries without bending stiffness are referred to as "cable catenaries" throughout this book.) These values include the horizontal tension component (H), top-end angle (θ_t), and the suspended length (s) and its horizontal projection (x). It is shown in chapter 12 that the values of the tensions and top-end angle (T_t, H, and θ_t) are effectively unchanged by the bending stiffness. Only the suspended length and its horizontal projection are increased, by an amount easy to evaluate.

In the data list, the SCR total length is in fact arbitrary. Nevertheless, it does allow the total horizontal projection, of the catenary plus seafloor flow line, to be calculated. Hence, changes in the top end offset position can be determined by comparing values of total horizontal projection given by different top tensions.

The file then determines the stresses that have to be checked in accordance with codes of practice. These include the von Mises' stresses, as well as the axial and circumferential membrane and bending stresses. As shown in chapter 4, all these stresses can be calculated directly at any point, from the effective tension, the curvature, and the internal and external pressures.

Bending will be a maximum near the TDP (but not precisely at the TDP, where it is theoretically zero). Likewise, there will also be some bending near the top end owing to pipe sag (but not precisely at the top end, where it is also theoretically zero unless there is an applied top-end moment).

Stresses have to be checked on internal and external surfaces, since the effect of bending is greatest on the external surface and the effect of pressure is greatest on the internal surface. Generally, the von Mises' stress is greatest on the internal surface. In *SCR-Example.xls,* the equations used to calculate all the different results are referenced in detail.

In chapter 15, a simple way of evaluating the natural periods of SCR *transverse* vibrations is presented. In *SCR-Example.xls,* approximate values of the natural periods for transverse modes between 1 and 15 are also calculated.

The preceding is given as an example to show the reader the potential use of the equations and expressions included in this book. It does not exhaust all that can be deduced about SCRs from the contents of this book.

Summary

This introductory chapter has discussed the principal different types of riser in use at the time of writing, with some history of their development. It has also given the reader an overview of the contents of the entire book, chapter by chapter and appendix by appendix.

To demonstrate the versatility of the equations and expressions derived in the book, they have been applied as an example to the preliminary analysis of an SCR in the Excel file *SCR-Example.xls,* which is referred to in several chapters.

Before proceeding to the detailed analysis of riser behavior, the first subject that must be clearly understood is the influence of internal and external pressures on riser stability and deflections. That is the subject of the next chapter.

References

[1]Guesnon, J., C. Gaillard, and E. Laval. 2000. A riser for ultra-deepwater drilling. *World Oil.* 221 (4), 90.

[2]Gelfgat, M., V. Tikhonov, B. Vygodsky, V. Chizhikov, and A. Adelman. 2006. New prospects in development of aluminum ultra deepwater risers. Paper presented at the Deep Offshore Technology Conference, Houston.

[3]Sparks, C. P. 1995. Riser technology for deepwater scientific drilling in the 21st century. Paper presented at the 14th Offshore Mechanics and Arctic Engineering Conference, Copenhagen.

[4]Utt, M., and Unocal. 2001. Extending drilling rig capabilities: Surface BOP stack applications in Southeast Asia. Paper presented to the American Association of Drilling Engineers, Sugar Land, Texas.

[5]Shanks, E., J. Schroeder, B. Ambrose, and R. Steddum. 2002. Surface BOP for deepwater moderate environment drilling operations from a floating drilling unit. Paper OTC 14265, presented at the Offshore Technology Conference, Houston;
Zang, J., E. Magne, D. Morrison, M. Efthymiou, C. Leach, and K. H. Lo. 2002. Pressured drilling riser design for drilling in ultradeep water with surface BOP. Paper presented at the UDET Conference, Brest, France.

[6]Fisher, E. A., and P. C. Berner. 1988. Non-integral production riser for Green Canyon block 29 development. Paper OTC 5846, presented at the Offshore Technology Conference, Houston.

[7]Neto, E., and J. Mauricio. 2001. Flexible pipe for ultra-deepwater applications: The Roncador experience. Paper OTC 13207, presented at the Offshore Technology Conference, Houston.

[8]Novitsky, A., and S. Serta. 2002. Flexible pipe in Brazilian ultra-deepwater fields: A proven solution. Paper presented at the Deep Offshore Technology Conference, New Orleans.

[9]Erb, P. R., C. L. Finch, and G. R. Manley. 1985. The Hutton TLP performance monitoring and verification program. Paper OTC 4951, presented at the Offshore Technology Conference, Houston.

[10]Phifer, E. H., F. Koop, R. C. Swanson, D. W. Allen, and C. G. Langer. 1994. Design and installation of Auger steel catenary risers. Paper OTC 7620, presented at the Offshore Technology Conference, Houston.

[11]Sparks, C. P. 1984. The influence of tension, pressure and weight on pipe and riser deformations and stresses. *Transactions of ASME. Journal of Energy Resources Technology.* 106 (March), 46–54.

2

Pipe and Riser Deflections and Global Stability: The Effective Tension Concept

The influence of tension, pressure, and weight on pipe and riser deflections and global stability has been treated by many authors.[1] Nevertheless, it is a subject that still causes difficulty and confusion for many.

Misunderstandings are widespread. Today accidents are fortunately infrequent, but expensive mistakes have been made in the past, not all of which have been the subject of publications.[2] The potential for incorrect design and operation of risers—and for incorrect orientation of research and development projects—is still enormous. Continued misunderstandings result in vast amounts of time being spent discussing the subject.

The approach used here closely follows one used in a previous publication.[3] It begins with a close look at Archimedes' famous law.

Archimedes' Law

Archimedes' law in its most general form states that when a body is wholly or partially immersed in a fluid, it experiences an upthrust equal to the weight of fluid displaced. This is illustrated in figure 2–1, in which a body is shown fully immersed in a fluid.

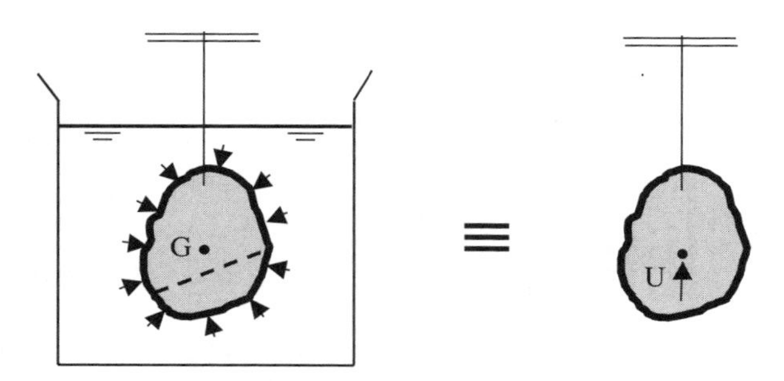

U = Upthrust = Weight of fluid displaced

Fig. 2–1. *Archimedes' law*

The argument taught to school children is that the pressure field is just able to maintain the displaced fluid in equilibrium, as shown in figure 2–2. Thus, it must provide an upthrust U equal to the weight of the fluid displaced W_f. Furthermore, since this upthrust can produce no rotation, it must act at the centroid of the displaced fluid, which is also the center of gravity G. Hence, it will also act at the centroid of the submerged body.

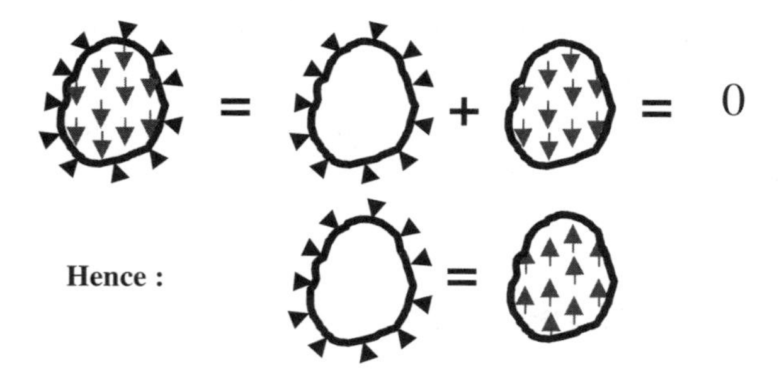

Fig. 2–2. *Pressure and weight acting in a fluid*

Thus, if the true weight of the body is W_t, the tension in the string will be given by the following, where $W_t - W_f$ is generally called the *apparent weight* W_a:

$$T = W_t - U = W_t - W_f \tag{2.1}$$

There are a number of important points to make about Archimedes' law:

- The law can be applied directly only to pressure fields that are completely closed. Note that for a suspended or floating body, the pressure field appears not to be closed; however, since the pressure at the surface is zero, the field can be considered to be closed.
- The law cannot be applied directly to *parts* of submerged bodies, such as that below the dotted line in figure 2–1.
- The law says nothing about internal forces or stresses.
- The closed pressure field, when combined with the distributed weight of the displaced fluid, can produce no resultant moment. The fluid would not be able to support the associated stresses.

Archimedes' Law— Proof by Superposition

Archimedes' law can also be deduced by *superposition*. This may be too abstract for school children, but it leads to the same results more clearly and directly. Since superposition will be used extensively in this book, it will be first used here to rederive Archimedes' Law.

In figure 2–3, the two systems shown (the submerged body and the displaced fluid) are both in equilibrium under the combined loads that include the effects of tension, pressure, and weight. Hence, if the two systems are superimposed and the forces on the displaced fluid are *subtracted* from those on the submerged body, the resulting equivalent system will also be in equilibrium.

Superposition of the two systems allows the identical pressure fields to be eliminated. All that remains in the resulting equivalent system is the tension T in the string and the apparent weight W_a, which is then simply the *difference* between the weights of the submerged body and the displaced fluid, as given by the following equation:

$$W_a = W_t - W_f \qquad (2.2)$$

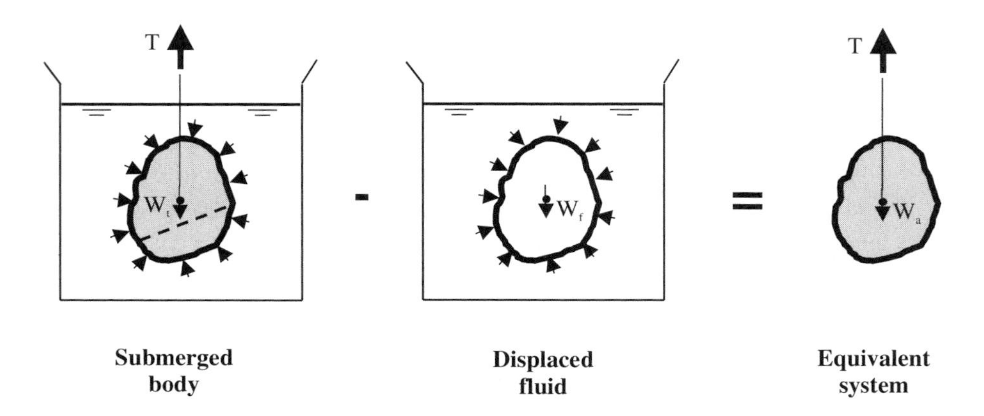

Fig. 2–3. *Archimedes' law by superposition*

Any two systems can be superimposed in this way. The only requirement is that they both be equilibrium. In the preceding, there is no need to specify that densities must be constant or that the upthrust acts at the centroid of one or other of the two systems. The argument can be applied directly to cases where the submerged body does not have a constant density; where the body is suspended across the interface between fluids of different densities, or where the density of the displaced fluid may vary vertically according to some law. As long as the displaced fluid segment represents exactly the fluid displaced by the submerged body, superposition can be used directly.

Internal Forces in a Submerged Body

In the calculation of the internal forces on a part of a submerged body, the problem is to take into account the pressure field that is not closed. Figure 2–4 shows the forces acting on the segment below the dotted line in figures 2–1 and 2–3. The resultant of the pressure field acting on the underside of the segment is unknown and cannot be determined directly using Archimedes' law.

Nevertheless, superposition allows the internal forces to be determined very simply. The middle sketch of figure 2–4 shows the forces acting on the displaced fluid segment including the *closed* pressure field. If these forces are subtracted from the forces on the body segment, the pressure field acting below the body is conveniently eliminated. However, the force

p_eA_e, owing to the pressure acting on the section, remains (where p_e is the pressure in the fluid and A_e is the cross-sectional area of the section). Since convention requires tension to be positive, this must be shown as a *tensile force*: $-p_eA_e$.

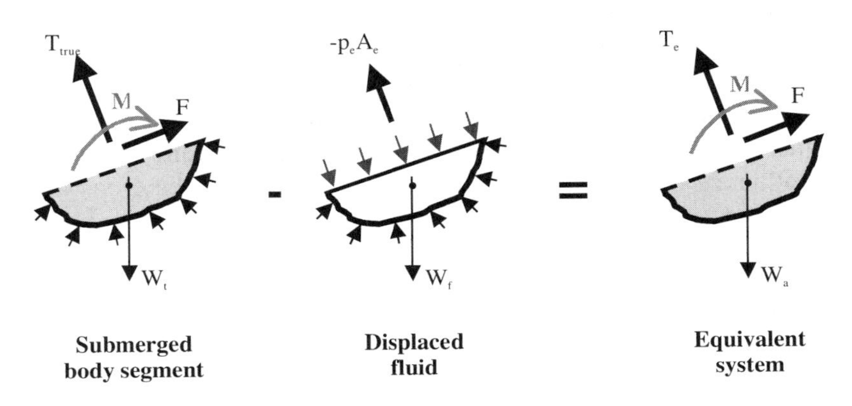

Fig. 2–4. *Internal forces acting on a submerged body segment*

The equivalent system (fig. 2–4, right-hand sketch) shows the resultant of the superposition. Once again, the apparent weight W_a is given by equation (2.2), where the weights W_t, W_f, and W_a correspond to the segment, rather than the whole body. Thus, the apparent weight W_a is in equilibrium with an effective tension T_e, a shear force *F,* and a moment *M,* which can be found by resolving forces normal and parallel to the section and by taking moments. The shear force *F* and the moment *M* are the same as on the body segment. (For the applications considered in this book, the minute moment created by the very slight pressure gradient across the section can be neglected.) The effective tension T_e is then related to the true tension T_{true} by

$$T_e = T_{true} - (-p_eA_e) = T_{true} + p_eA_e \tag{2.3}$$

According to convention, tensile forces are positive. However, according to a further convention, pressures are also positive. The positive sign in the right-hand side of equation (2.3) results from the contradiction between the two conventions. The effective tension T_e is nevertheless the *difference* between the *tensions* acting on the body segment and the displaced fluid segment, just as the apparent weight is the difference between their weights.

Curvature, Deflections, and Stability of Pipes and Risers under Pressure

The preceding arguments can be extended to the case of pipes and risers under pressure. Figure 2–5 shows equivalent force systems for the case of a pipe subjected only to internal pressure p_i. For clarity, moments and shear forces have been omitted, but that does not influence the argument. A pipe segment of length δs is shown curved and in equilibrium under the combined influence of pipe weight, internal pressure, and the true wall tension T_{tw} acting in the pipe wall.

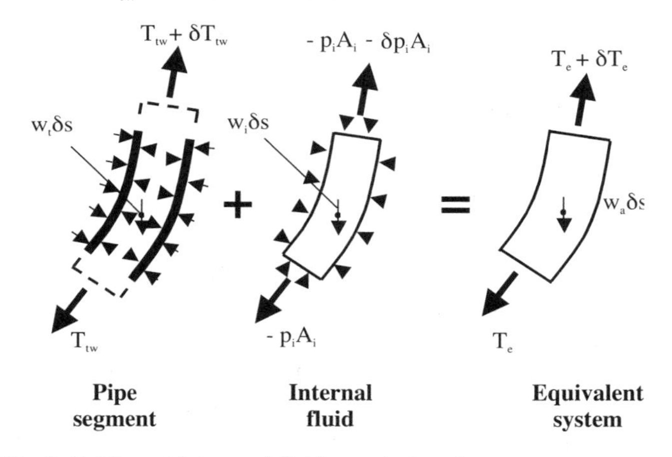

Fig. 2–5. *Pipe with internal fluid—equivalent force systems*

The pressure field acting on the internal fluid column is closed and in equilibrium with the weight of the internal fluid. The lateral pressures acting on the pipe wall are equal and opposite to those acting on the internal fluid. Hence, by superposition and *addition* of the two force systems, those lateral pressures are eliminated. However, the axial "tension" in the fluid column $-p_iA_i$ remains (where p_i is the internal pressure and A_i is the internal cross-sectional area of the pipe). This leads to the equations for the effective tension T_e and apparent weight w_a of the equivalent system:

$$T_e = T_{tw} + (-p_iA_i) \tag{2.4}$$

$$w_a = w_t + w_i \tag{2.5}$$

When external pressure p_e is also present, the same approach can still be used, as shown in figure 2–6. By the addition of the force systems acting on the pipe segment and the internal fluid and then the subtraction of the force system acting on the displaced fluid, all lateral pressure effects are eliminated. In figure 2–6, w_t, w_i, w_e, and w_a are the weights per unit length of the tube, the internal fluid column, the displaced fluid column, and the equivalent system, respectively.

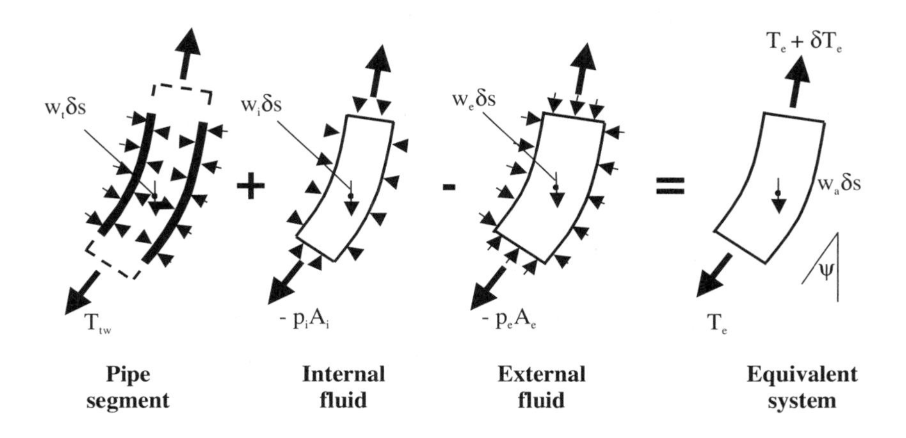

Fig. 2–6. *Pipe with internal and external fluids—equivalent force systems*

The equations for the effective tension T_e and the apparent weight w_a then become

$$T_e = T_{tw} + (-p_i A_i) - (-p_e A_e) \tag{2.6}$$

$$w_a = w_t + w_i - w_e \tag{2.7}$$

Furthermore, the two concepts are related, as can be seen from the right-hand sketch in figure 2–6. For an element of length δs, resolution of forces in the axial direction gives

$$\frac{dT_e}{ds} = w_a \cos\psi \tag{2.8}$$

which for small angles with the vertical becomes $dT_e/ds = dT_e/dx = w_a$.

Since for any fluid the combined effects of its weight and enclosed pressure field can produce no resultant moment anywhere (see the fourth bullet point following equation [2.1]), the *bending effects* of forces on the equivalent system are precisely the same as those on the pipe segment. Therefore, the simplest way to take into account the effects of internal and external pressure on pipe or riser curvature, deflection, and stability is to use effective tension and apparent weight in the corresponding tensioned-beam calculations.

The effective tension, at any point along a riser, can be obtained most simply by considering the equilibrium of the segment between the point and the riser top end, taking into account the riser top tension and the segment apparent weight. The true wall tension T_{tw} can then be found from equation (2.6).

Effective Tension—a Physical Interpretation/Definition

The interpretation that can be given to the effective tension equation depends on how it is derived. Those who choose to integrate the effect of pressures on a perfectly cylindrical elastic tube undergoing small deflection will also arrive at equation (2.6). However, they will not immediately be able to deduce the effect of those pressures on tubes of other geometries, made of nonelastic materials, subject to large deflections.

The arguments used in this chapter to derive equation (2.6) are very general. The only limitation is that each of the component systems shown in figure 2–6 must be in static equilibrium. The cross-section of the tube (deliberately not defined as $\pi D^2/4$) is not necessarily circular. Likewise, the argument is not limited to tubes of uniform cross-section. The material is not necessarily of constant density, nor is it necessarily elastic. Furthermore, no mention has been made of small- or large-angle deflections. The equation as derived is therefore of completely general validity.

Although they would be difficult to draw, sketches similar to figure 2-6 could be made for more complicated cases, including several pipes connected together, with some pipes within others. This would lead to the general equation for effective tension:

$$T_e = \sum T_{tw} + \sum(-p_i A_i) - \sum(-p_e A_e) \qquad (2.9)$$

from which the following physical interpretation can be deduced: *Effective tension is the total axial force in the pipe/riser column, including internal fluid columns, less the axial force in the displaced fluid column (tension positive).* The axial force in the displaced fluid column can be considered to be a *datum force* to which the axial force in the riser plus contents must be referred.

The general equation for apparent weight, corresponding to equation (2.7), is

$$w_a = \sum w_t + \sum w_i - \sum w_e \qquad (2.10)$$

Comparison between equations (2.9) and (2.10) shows that there is a precise term-by-term parallel between effective tension and apparent weight. That is already clear from figure 2–6. Yet apparent weight has never caused difficulty or confusion for anyone. No one has ever objected to the statement that "apparent weight is the sum of the weight of the pipe/riser plus contents less the weight of displaced fluid." No one has ever argued that apparent weight is just a mathematical concept in a differential equation.

Effective Tension— a Mathematical Approach

In contrast to apparent weight mentioned previously, some commentators have objected that "effective tension is simply a grouping of forcelike terms in the tensioned-beam differential equation, with no physical significance." They have argued that the first term of equation (2.6) is an *axial effect* and the other two terms are *lateral effects*. Given this objection, it may be helpful to rederive equation (2.6) mathematically, showing how such a derivation can also lead to the same physical interpretation of effective tension given in italics below equation (2.9).

The governing differential equation for a tensioned beam is derived in appendix A. For a riser made from elastic materials, with uniform bending stiffness, undergoing small deflections,

$$EI\frac{d^4y}{dx^4}-\frac{d}{dx}\left(T\frac{dy}{dx}\right)-f(x)=0 \qquad (2.11)$$

where EI is the bending stiffness, T is the tension, and $f(x)$ is the lateral load per unit length; x is the vertical axis, and y is the horizontal axis.

It is important to realize that *all* terms of equation (2.11) are differentials of lateral forces, with respect to x. The first term is the rate of change of the shear force, the second term is the rate of change of the lateral component of axial tension, and the third term is the external lateral load per unit length. This can be clearly understood from figure A–1 of appendix A.

If two or more beams are coupled together, equation (2.11) can be written individually for each beam, providing that the interaction forces between them are included. For simplicity, it is assumed that there is no axial adhesion between the beams. Axial adhesion would complicate but not fundamentally alter the argument. Hence, if there are just two beams (subscripts 1 and 2), their paired equations become

$$\begin{aligned}(EI)_1\frac{d^4y}{dx^4}-\frac{d}{dx}\left(T_1\frac{dy}{dx}\right)&=f(x)_1+f_{i1}\\(EI)_2\frac{d^4y}{dx^4}-\frac{d}{dx}\left(T_2\frac{dy}{dx}\right)&=f(x)_2+f_{i2}\end{aligned} \qquad (2.12)$$

where f_{i1} and f_{i2} are equal and opposite interaction forces between the beams.

Since the beams are coupled together, they have the same deflected shape. Thus, the equations can be summed, which leads to the elimination of the interaction forces f_{i1} and f_{i2}:

$$\left(\sum EI\right)\frac{d^4y}{dx^4}-\frac{d}{dx}\left[\left(\sum T\right)\frac{dy}{dx}\right]=\sum f(x) \qquad (2.13)$$

When deriving equation (2.13), no assumption was made about the relative stiffness or axial tension contribution of the individual beams. It is not limited to beams with both bending stiffness EI and axial tension T. A tensioned cable running through guides on a riser would contribute to the tension $\sum T$ but not to the stiffness $\sum EI$. Likewise, a beam with stiffness EI but zero tension would only contribute to $\sum EI$. A column under axial *compression* would contribute to $\sum EI$ but reduce the total tension $\sum T$.

The *internal fluid column* is the limiting case of a column under axial compression p_iA_i but with zero bending stiffness. Hence, it contributes to (reduces) $\sum T$ without contributing to $\sum EI$. It can of course do this only by being enclosed in another component of the system, namely the riser. For a riser with an internal fluid column, equation (2.13) becomes

$$EI\frac{d^4y}{dx^4} - \frac{d}{dx}\left[(T_{tw} - p_iA_i)\frac{dy}{dx}\right] = f(x) \qquad (2.14)$$

where the term $(T_{tw} - p_iA_i)$ is seen to be the total axial force in the *riser plus internal fluid column,* which contributes $(T_{tw} - p_iA_i)\, dy\, /\, dx$ to the lateral force balance of equation (2.14). This is coherent with the physical interpretation of effective tension given in italics following equation (2.9).

If external pressure is present, then the *lateral load* f_e due to external pressure will be the same for the riser as for the displaced fluid column. The paired equations for the riser plus contents and the displaced fluid column can be written as

$$\begin{aligned} EI\frac{d^4y}{dx^4} - \frac{d}{dx}\left[(T_{tw} - p_iA_i)\frac{dy}{dx}\right] &= f(x) + f_e \\ -\frac{d}{dx}\left[(-p_iA_i)\frac{dy}{dx}\right] &= f_e \end{aligned} \qquad (2.15)$$

Subtracting the second equation from the first eliminates f_e, leading to

$$EI\frac{d^4y}{dx^4} - \frac{d}{dx}\left[(T_{tw} - p_iA_i + p_eA_e)\frac{dy}{dx}\right] = f(x) \qquad (2.16)$$

The term enclosed in parentheses in the second term of equation (2.16) is the effective tension as given by equation (2.6). In deriving equation (2.16), all the components of this term were seen to be *axial forces*. Therefore, there is no contradiction between the effective tension, derived mathematically, and the physical interpretation of effective tension given in italics following equation (2.9).

With substitution for the effective tension from equation (2.6), equation (2.16) becomes

$$EI\frac{d^4y}{dx^4}-\frac{d}{dx}\left(T_e\frac{dy}{dx}\right)=f(x) \qquad (2.17)$$

which for small-angle deflections, where $dT_e/dx = w_a$, becomes

$$EI\frac{d^4y}{dx^4}-T_e\frac{d^2y}{d^2x}-w_a\frac{dy}{dx}=f(x) \qquad (2.18)$$

Comparisons with Analogous Engineering Concepts

Since the influence of pressure on pipe/riser curvature and stability causes such difficulty in the offshore industry, it may be helpful to examine how similar problems are treated in other industries. For example, civil engineers face similar problems with prestressed concrete beams and composite steel/concrete columns.

Figure 2–7*a* shows an example of a prestressed concrete beam, in which steel wires or cables, parallel to the beam axis, are highly stressed in tension before being locked off to the concrete. This results in very high axial compressive loads in the concrete. It is true that civil engineers when first confronted with the subject tend to be alarmed that the huge axial compressive loads in the concrete will cause slender beams to buckle. However, they quickly observe that they do not! They then accept without difficulty that the axial compression in the concrete is exactly balanced by the tension in the wires/cables. Therefore, the total axial force in the beam is zero. Hence, there can be no tendency to buckle.

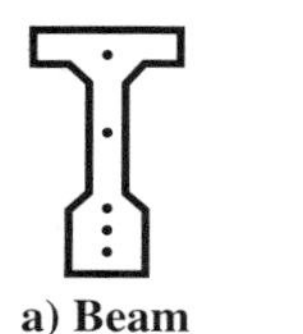

Fig. 2–7. *Example beam and column sections from civil engineering*

Civil engineers are also frequent users of composite steel/concrete columns. Figure 2–7*b* shows an example consisting of a cylindrical steel caisson filled with concrete. Figure 2–8 shows forces acting on a short length of such a column (moments and shear forces are omitted for clarity). The forces closely resemble those of figure 2–5.

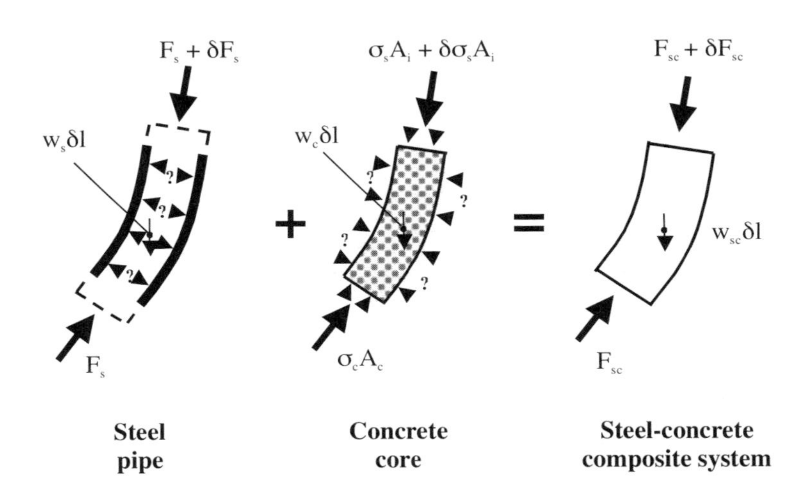

Fig. 2–8. *Analysis of a composite steel-concrete column*

In the study of column stability, it is standard practice among civil engineers to analyze the combined steel/concrete column. This has the advantage of eliminating the interaction stresses between the steel and the concrete, which are not known precisely.

An engineer could choose to analyze one component of such a column (say the steel cylinder) and deduce the *effect* of the interaction stresses by analyzing the equilibrium of the second component (the concrete core). He would conclude that the steel cylinder behaves as if it were subject to *effective forces*: an *effective axial load,* equal to the sum of the axial forces in the two components; an *effective shear force,* equal to the sum of the shear forces; and an *effective moment,* equal to the sum of the moments. Also, the cylinder behaves as if it has an *effective bending stiffness,* equal to the sum of the stiffnesses of the two components. (In case some readers object to this last statement, the bending stiffness of a system comprising *concentric* cylinders is equal to the sum of the stiffnesses of the cylinders, even if there is adhesion between them. For other configurations, the total stiffness can be much greater.) Thus, the *equivalent system* analyzed would have all the characteristics of the *combined system*. When transposed to the pipe/riser system, all forces and characteristics of the internal fluid column are zero, except for the axial load. Hence, only the axial forces have to be summed, as stated in italics following equation (2.9).

Another example can be taken from hydraulic engineering. The interpretation of effective tension, given in italics following equation (2.9), requires the internal fluid column to be considered as part of the riser system. Although some offshore engineers are reluctant to do this, hydraulic engineers have no such reluctance.

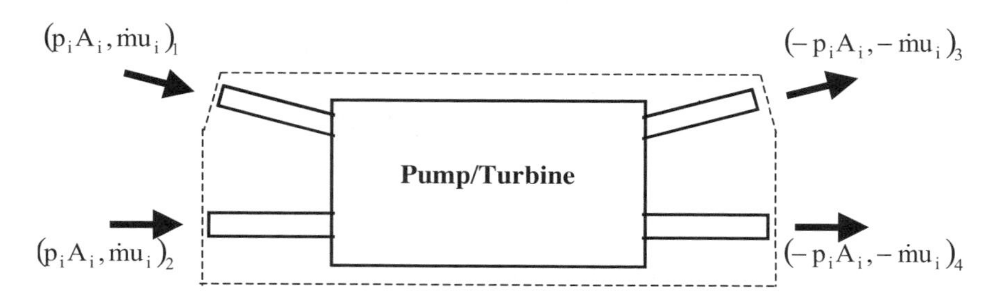

Fig. 2–9. *Forces acting on a pump or turbine*

When calculating the forces acting on the foundations of a pump or turbine (see fig. 2–9), a hydraulic engineer could integrate the effects of pressure acting on every nook and cranny inside the installation, to find the resultant. However, the same result can be obtained much more simply by treating the fluid as part of the system and by considering the balance of pressure-force vectors pA and the rate-of-change-of-momentum vectors $\dot{m}u_i$ of the fluids entering and leaving the system.

Requirements of Codes of Practice

All codes of practice require the global behavior of pipes and risers to be calculated using effective tension. This is generally defined in one of two ways:

- Some codes quote equation (2.6), but without parentheses, as

$$T_e = T_{tw} - p_i A_i + p_e A_e \qquad (2.19)$$

- Some codes mention that effective tension is the axial tension calculated at any point of the riser by considering only the top tension and the apparent weight of the intervening riser segment.

Both these definitions have already been given in this chapter. However, when they are compared, they appear to be in contradiction. The first definition includes pressure terms, whereas the second does not. It may be obvious to the reader why there is actually no contradiction between them; if not, figure 2–10 may be helpful.

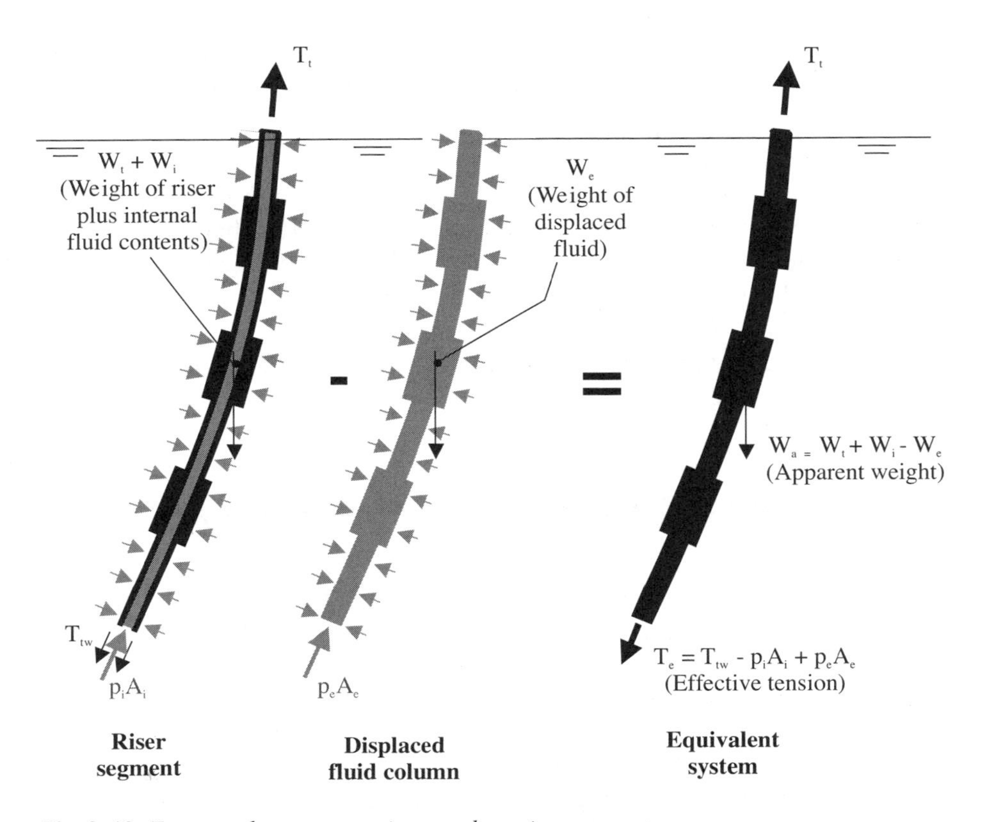

Fig. 2–10. *Forces and pressures acting on a long riser segment*

Figure 2–10 is similar to figure 2–6 but drawn for the whole upper segment of riser above some section of interest. The left hand sketch of Figure 2–10 shows the forces that maintain the upper segment in equilibrium. These include the top tension, the true weight (in air) of the pipe plus internal fluid, the external pressure field, the true wall tension T_{tw}, and the thrust in the internal fluid column $p_i A_i$ at the section of interest. The shear force and moment at the section are omitted for clarity.

The central sketch of Figure 2–10 shows the forces that maintain the displaced fluid column in equilibrium, including the axial thrust $p_e A_e$ at the lower end. Superimposing the two force systems and subtracting the one from the other eliminates the external pressure field. Collecting terms as shown in the right hand sketch of Figure 2–10 leads once again to the familiar equations for effective tension and apparent weight. It also clearly shows that the effective tension T_e is in equilibrium with the top tension and the apparent weight, in agreement with the second definition.

In the left hand sketch of figure 2–10, if the top tension, the apparent weight, and the external pressure field are maintained constant, then the sum of the forces ($T_{tw} - p_i A_i$) at the segment bottom end must also remain constant. Thus, the riddle of the apparent contradiction between the two definitions, given at the beginning of this section, is explained. *The effective tension (T_e) is independent of pressure (depending only on top tension and apparent weight). The true wall tension (T_{tw}) varies with pressure.* Readers who have difficulty in accepting the preceding may find the discussion in appendix B helpful.

Excel File *Riser-Tensions.xls*

The Excel file *Riser-Tensions.xls* allows the reader to calculate the effective tension at the lower end of a segment of riser, similar to that shown in figure 2–10 but perfectly uniform and vertical (see fig. B–1*c* and *d* of appendix B). Results calculated using the two methods given previously (by equation [2.19] and the bullet point following it) are compared for capped and uncapped risers. The reader can use the file with his or her own data. For ease of printing, a further version of the file, named *Riser-Tensions (print).xls,* is provided.

In the first method, the tension in the pipe wall (T_{tw}) is found by analyzing the vertical forces acting on the pipe wall. Equation (2.19) is then used to find the effective tension (T_e).

In the second method, the effective tension (T_e) is found from the top tension and the apparent weight of the pipe segment between the top end and the section of interest. Equation (2.19) is then used to work back to the true wall tension (T_{tw}). The file shows clearly that

- The two calculation methods give the same results
- The effective tension is the same for both the capped and the uncapped tubes
- Changes in the internal pressure of the capped tube influence the true wall tension (T_{tw}) but leave the effective tension (T_e) unchanged

For the very simple examples treated by *Riser-Tensions.xls,* both methods are straightforward. However, for more complicated cases, it is easier to find the effective tension first from the apparent weight, as given by the bullet point following equation (2.19), and to then use equation (2.19) to work back to the true wall tension.

Summary

This chapter has shown that effects of internal and external pressures on pipe and riser bending and stability are correctly taken into account by using the concepts of effective tension and apparent weight. The effective tension equation has been derived by two different ways, using physical and mathematical arguments.

Effective tension has been given a physical interpretation, which involves considering the pipe wall and the internal fluid column as parts of a combined system and the displaced fluid column as a datum system. All terms in the effective tension equation are interpreted as axial forces, in the pipe wall and fluid columns. Effective tension is then the axial force in the combined system less the axial force in the datum system.

Analysis of similar combined systems is common practice in civil engineering. Hydraulic engineers are accustomed to considering axial forces in fluid columns.

The requirements of codes of practices have been discussed, along with their different ways of defining effective tension.

The Excel file *Riser-Tensions.xls* has been presented. The file can be used to demonstrate that changes in internal pressure in a capped tube have no influence on effective tension, but they do influence the true wall tension.

The interpretation of effective tension, in terms of axial forces in the pipe wall and fluid columns, can help elucidate many particular pipe and riser problems that might otherwise cause confusion. Several such problems are discussed in the next chapter.

References

[1]Klinkenberg, A. 1951. Neutral zones in drill pipe and casing and their significance in relation to buckling and collapse. *Drilling and Production Practice, API.* 64–79.

Lubinski, A. 1975. Influence of neutral axial stress on yield and collapse of pipe. *Journal of Engineering for Industry, ASME.* May, 400–407.

Morgan, G. W. 1977. Analyzing top tension. *Ocean Resources Engineering.* February, 40–50, and April, 12–24.

Lubinski, A. 1977. Necessary tension in marine risers. *Revue de l'Institut Français du Pétrole.* Vol. XXX11, No. 2, 233–256, and 32 (6), 873–895.

Goins, W. C. 1980. Better understanding prevents tubular buckling problems. *World Oil.* January, 101–106, and February, 35–40.

Pattillo, P. D., and B. V. Randall. 1980. Two unresolved problems in well bore hydrostatics. *Petroleum Engineer International.* July, 24–32.

Bournazel, C. 1980. Conséquences de l'effet de fond sur l'instabilité des conduites soumises à une pression interne. *Revue de l'Institut Français du Pétrole.* 35 (1), 79–100.

Bernitsas, M. M. 1980. Riser top tension and riser buckling loads. Paper presented at the ASME Winter Annual Meeting, Chicago.

McIver, D. B., and R. J. Olson. 1981. Effective tension—now you see it, now you don't. Paper presented at the 37th Mechanical Engineering Workshop and Conference, Petroleum Division ASME, Dallas.

Chakrabarti, S. K., and R. E. Frampton. 1982. Review of riser analysis techniques. *Applied Ocean Research.* 4 (2), 73–90.

Sparks, C. P. 1984. The influence of tension, pressure and weight on pipe and riser deformations and stresses. *Transactions of ASME. Journal of Energy Resources Technology.* 106 (March), 46–54.

[2]Tilbe, J. R., and G. Van de Horst. 1975. Risers: Key cause to North Sea down time. *Oil & Gas Journal.* March 10th, 71–73.

[3]Sparks, C. P. 1984. The influence of tension, pressure and weight on pipe and riser deformations and stresses. *Transactions of ASME. Journal of Energy Resources Technology*. 106, 46–54.

3 Application of Effective Tension: Frequent Difficulties and Particular Cases

This chapter is devoted to discussing particular difficulties that recur frequently in connection with effective tension. The concepts of *end loads* and *end effects,* as well as their application and significance to risers, are first examined. A number of particular questions and applications are then discussed, some of which have already been given summary treatment previously.[1]

The special case of multi-tube risers, including tubes within tubes, is then considered. In particular, a way of decomposing the components of effective tension and apparent weight of a multi-tube riser is presented.

The application of effective tension to riser dynamic cases is then discussed. The chapter closes with a list summarizing reasons for frequent confusion about effective tension.

End Loads and End Effects

When discussing the effect of pressure on pipe behavior, many engineers and technicians polarize their thoughts on the pipe end loads or end effects (sometimes called "end cap loads" or "end cap effects"). These concepts are helpful in understanding certain cases but can increase confusion when applied to others.

Horizontal tubes

Figure 3–1 shows three horizontal capped tubes subject to internal pressure p_i. Tube *a* is straight and of uniform diameter. Tube *b* is straight but of nonconstant diameter. Tube *c* is not straight.

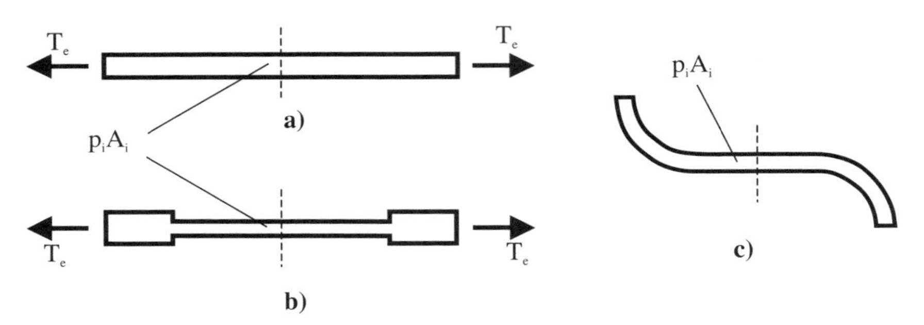

Fig. 3–1. *Three horizontal capped tubes under internal pressure*

End load is helpful in understanding the forces in tube *a*. If the tube is imagined cut along the dotted line, then plainly the force in the tube wall T_{tw} will be equal to the sum of the pressure end load p_iA_i plus the applied axial load T_e:

$$T_{tw} = p_iA_i + T_e \qquad (3.1)$$

The subscript "e" for the applied tensile load is convenient since T_e can be interpreted in three ways:

- As an external tension (i.e., external to the tube)
- As an excess tube wall tension (i.e., in excess of p_iA_i)
- As the tube effective tension (see equation [2.19])

These interpretations are unchanged when external pressure is added. T_e is then given as follows:

$$T_{tw} = p_iA_i - p_eA_e + T_e \qquad (3.2)$$

T_e remains an external tension, an excess wall tension (in excess of $p_iA_i - p_eA_e$), and the effective tension.

For the other two tubes, *end load* is unhelpful for calculating internal forces because of the changes of diameter and the bends. If instead *end effect* is taken to mean the integral of all pressure-induced axial forces between the section under consideration and the tube end, then it will again give $(p_iA_i - p_eA_e)$ for each tube, where A_i and A_e are the respective internal and external cross-sectional areas at the point under consideration.

In reality, it is impossible to carry out the preceding integral for complicated cases. For example, the ends of the trans-Siberian gas pipeline are separated by thousands of kilometers. In Siberia, the ends are located deep in the earth in a multitude of wells. In Europe, they consist of literally millions of ends in domestic homes! Fortunately, it is not necessary to actually carry out the integral, since the result always gives $(p_iA_i - p_eA_e)$ — that is, the axial force in the internal fluid column less the axial force in the displaced fluid column, at the section concerned.

Nonhorizontal tubes

End loads and end effects can also be helpful in understanding the effect of pressures on perfectly vertical uniform tubes, as explained in appendix B. However, the situation becomes much more confusing for risers, which are neither perfectly vertical nor horizontal. Figure 3–2 shows five near-vertical tubes, all subject to the same effective tension T_e.

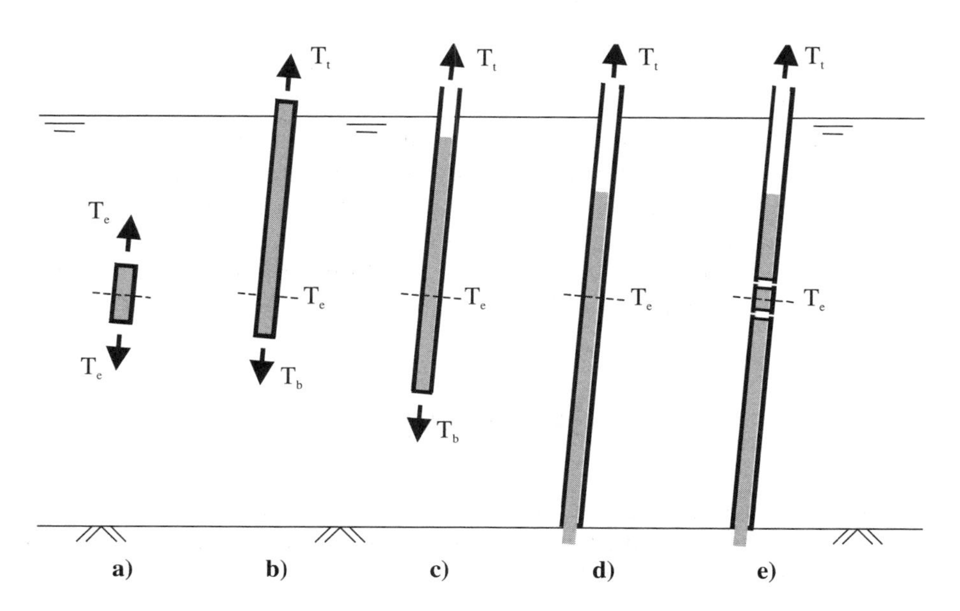

Fig. 3–2. *Five near-vertical tubes under pressure*

Tube *a* of figure 3–2 is simply tube *a* of figure 3–1 reorientated. If the tube is short and tensions and pressures are high, the self weight may be neglected. In that case, consideration of pressure-induced end loads returns to equation (3.2), which leads to the same triple interpretation of T_e (as an external tension, an excess tension, and the effective tension) as given prior to equation (3.2).

Tube *b* is long and self weight can no longer be neglected. The pressure end loads are different. The weight of the pipe wall and the internal fluid column must be included in the calculation to get back to the tension equation at the section under consideration. This yields the effective tension, as given earlier by equation (2.19), with p_iA_i and p_eA_e at the section concerned.

Tube *c* has only one closed end. Hence, there is an end load in the lower part but not in the upper part.

Tube *d* is full of fluid and is open at both ends. Thus, it has no end load.

Tube *e* is exactly the same as tube *d* but with double diaphragms added just above and below the section of interest. The external fluid is allowed to invade the short sections between each pair of diaphragms. Thus, the section of interest of tube *e* can be considered to be part of a tube with end loads or end effects.

Hence, if double diaphragms are imagined either side of the section of interest, $(p_iA_i - p_eA_e)$ can still be considered to be an end load or an end effect, and the same triple interpretation of T_e given prior to equation (3.2) can be maintained.

The effective tension equation is the same for each tube in figure 3–2. Interpretation of the equation in terms of end effects or end loads is therefore always possible, but it requires imagination. It is generally simpler to return to the physical interpretation, given in italics following equation (2.9), and to accept p_iA_i as the axial force in the internal fluid column and p_eA_e as the axial force in the displaced fluid column at the section under consideration.

Buoyancy

In the discussion of riser behavior, considerable confusion can be generated by debate about *buoyancy,* which is a concept often referred to loosely in the offshore industry.

The Archimedes upthrust, or buoyancy, acting on a submerged body was recalled at the beginning of chapter 2 (see figs. 2–1 to 2–3). It is a *volumetric* force in the sense that it is the resultant of the closed pressure field acting on the enclosed volume. It is equal to the weight of fluid displaced by the body and acts at the centroid of the submerged volume (see point G of fig. 2–1). The concept can be applied to any fully submerged body. It can also be applied to any suspended or floating body—such as a suspended riser or a ship hull—for which the pressure fields can be considered to be closed (see the first bullet point following equation [2.1]). For a suspended riser, the buoyancy is equal to the weight of fluid displaced, which for a vertical riser of uniform section is equal to the pressure × area ($p_e A_e$) acting at the riser lower end. Note, however, that the buoyancy force acts at the centroid of the *submerged volume,* at the midheight of the submerged length (see point G of fig. 2–1), not at the riser lower end. Ships would capsize if buoyancy acted at the keel level instead of at the centroid of their displaced volume.

Confusion arises when discussing the buoyancy of *part* of a submerged object (see fig. 2–4), such as a segment of riser, since it is subject to a pressure field that is *not* closed. The confusion is particularly flagrant if the riser pipe concerned is vertical and of uniform section. The wall of such a riser is continuous. Hence, the fluid pressures will act only horizontally on the segment and will have no vertical component. It is tempting to say that such a riser segment has no buoyancy. Since the segment can be positioned anywhere along the riser length, that would imply that the entire riser has no buoyancy, except at the surface at the lower end. That plainly does not agree with Archimedes' conception of buoyancy as an upthrust acting on a submerged volume.

The confusion is further increased when considering the stability of a vertical uniform riser connected to the seabed, since the external fluid pressures then do not even apply a vertical force to the surface at the riser lower end! Yet if the riser has *negative* apparent weight (i.e., is lighter than water), it will remain vertical and stable even if the top tension is reduced to zero. How does it do that without collapsing in a heap on the seabed? Some would argue that if the riser did depart from the vertical, forces with vertical components would be generated which would return it to the vertical.

The aforementioned confusion results from trying to apply the concept of buoyancy directly to pressure fields that are not closed. Since such pressure fields enclose no volume and hence have no centroid, Archimedes' law cannot be applied to them directly. They can be analyzed by artificially closing the fields and taking into account all the resultant forces, as was done in chapter 2 (see figs. 2–4 to 2–6). That led to the concepts of effective tension and apparent weight. The notion of buoyancy is already integrated into those concepts and requires no separate consideration. The effects of pressure on riser *stability* and *deflected shape* are always taken into account simply and accurately by considering effective tension and apparent weight whatever the orientation and configuration of the riser.

Recurrent Questions and Problems

There are a number of recurring questions and problems relating to the effects of pressure. It may be useful to discuss these in detail. Further questions, involving stresses and axial strains, are mentioned in chapters 4 and 5.

Lateral loads resulting from axial forces in fluid columns

It has been shown in chapter 2 that the curvature, deflection, and stability of a pipe under *internal* pressure depend on the effective tension, which is the sum of the axial forces in the pipe wall and the internal fluid column. Some engineers and technicians have great difficulty in understanding how an axial force in the internal fluid column can have any influence whatever on pipe curvature, deflection, and stability, arguing that "the force passes straight through the pipe from one end to the other without even adhering to the pipe wall"!

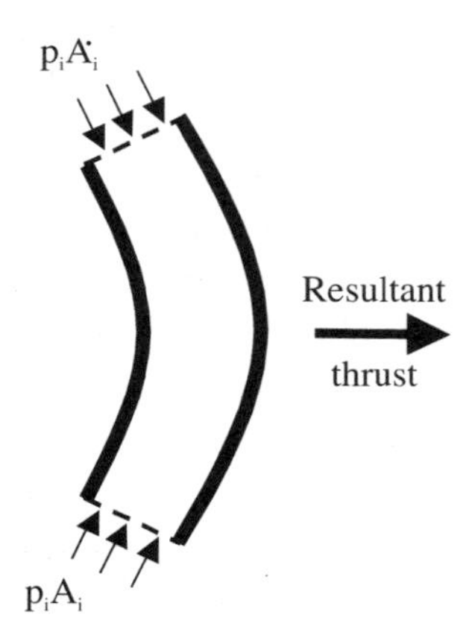

Fig. 3–3. *Lateral thrust resulting from axial load in fluid column*

Figure 3–3 shows a curved pipe subjected to internal pressure. The axial forces in the fluid column can be seen to provide a lateral thrust component that is transmitted to the pipe. Furthermore, the thrust plainly increases with pipe curvature, which increases with pipe deflection. Hence, the thrust is a destabilizing force.

If the reader is tempted to object that such a lateral thrust can exist only if the pipe is not straight, he or she should assume that there is always some small initial curvature. The existence of a small initial perturbation is common and is a necessary assumption in the analysis of many stability problems.

Buckling of suspended rods, pipes, and cables

Many engineers have wondered whether the axial compressive force, resulting from hydrostatic pressure, at the lower end of a suspended string (pipe, rod, or cable) will eventually cause the string to buckle if the pressure is great enough. This question needs to be taken seriously since it has perversely influenced some riser projects in the past. At least one article has been entirely devoted to the subject, and divers have even been sent down to try and clarify what really happens.[2]

Laypeople—and in particular, fishers and sailors—know the answer to this problem. They know that their lines will never buckle however deep they descend, even though they may not know why!

Those who are troubled by this problem tend to use the following argument: "The lower end of a suspended string is subjected to an axial compression $p_e A_e$ that increases with depth. Pressure effects above the lower end are lateral, equal, and opposite and therefore cancel out. Therefore, the string will eventually buckle when the axial compression becomes large enough." This is a *false argument,* as will be demonstrated in the following paragraphs.

Figure 3–4*a* shows a suspended string and its displaced fluid column, both of which are subject to an axial compressive force $p_e A_e$. The fact that the string will never buckle can be argued in several ways:

- First, the arguments of the earlier sections of this book can be applied. Namely, buckling stability is governed by effective tension, which can be found from the top tension and the apparent weight. If the apparent weight is positive, the effective tension will be zero at the bottom end and positive at all points above it. Thus, the string will never buckle.

- Second, the stability of the displaced fluid column can be considered. The displaced fluid column is also subject to the large axial force p_eA_e, yet it remains in stable equilibrium even though it has no resistance to anything (except hydrostatic pressure). Plainly, any string will also remain in equilibrium without buckling, providing that it is subject to a lesser axial compressive force than the displaced fluid column. This will always be the case if the apparent weight is positive.
- Third, there is a flaw in the false argument: lateral pressures are *not* equal and opposite and do *not* cancel out. Figure 3–4*b* shows a slightly curved segment of the displaced fluid column in equilibrium under the applied pressures. As can be seen in this sketch, the lateral pressures provide a net thrust that is exactly balanced by the lateral component of the axial compressive loads p_eA_e. The submerged string is subject to the same lateral thrust. If the axial compressive force is *less* in the string than in the displaced fluid column (p_eA_e), the lateral thrust will provide a net restoring force. The string will then straighten and never buckle. This is always the case if the apparent weight is positive.

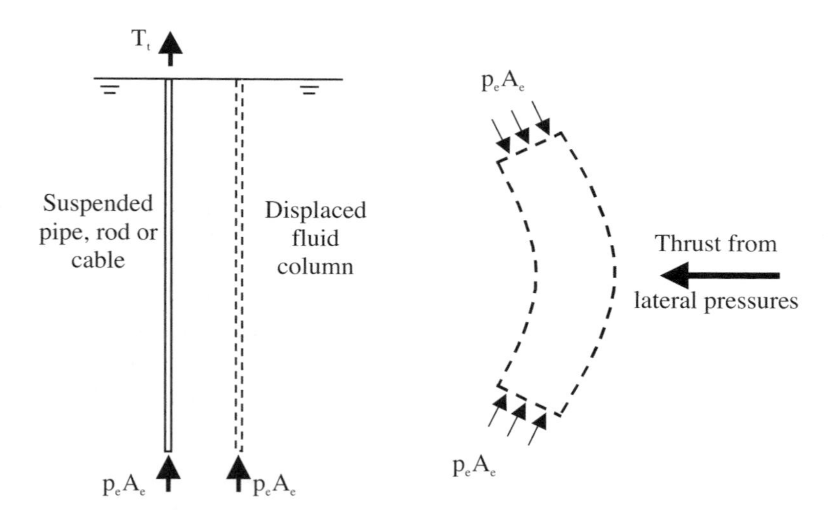

Fig. 3–4. *Stability of suspended strings*

The axial compression in the string will be only a destabilizing force if it is greater than the force $p_e A_e$ in the displaced fluid column. That will be true only if the apparent weight is negative, in which case the string will tend to float. This destabilizing force can be felt when a light stick (lighter than water) is plunged vertically into water. It quickly becomes impossible to keep the stick vertical because of the destabilizing force.

Buckling of pressurized pipes

Figure 3–5 shows two identical pipes under an axial compressive load *P*. Tube *a* is pressurized, while tube *b* is not. It is tempting to think that the pressure end load $p_i A_i$ in tube *a* will at least partially resist the external load *P* and that tube *a* should therefore resist a higher axial load before buckling than tube *b*. In fact, this is not so.

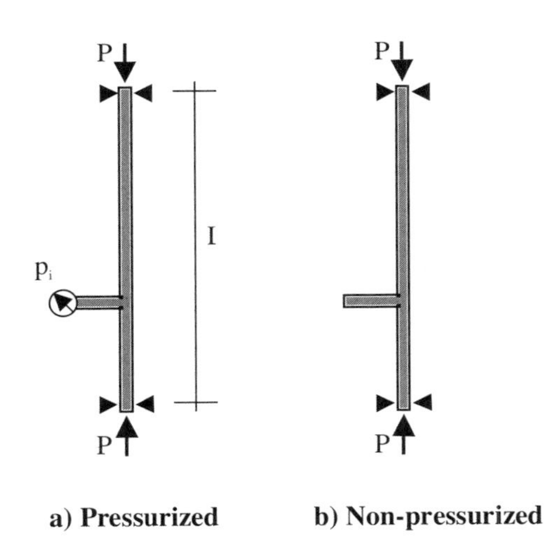

Fig. 3–5. *Buckling of pipes with and without internal pressure*

The *effective axial force* in each column is the same, equal to *P*. The pipes will both buckle when that axial force reaches the Euler buckling load: $P_E = \pi^2(EI)/L^2$. The way in which the effective axial force *P* is shared between the pipe wall and the internal fluid column has no influence on buckling.

Influence of pressure end load on stability

Some engineers are tempted to believe that pressure end load can contribute to holding up the riser top end, as though it were an alternative or an addition to *top tension*. This is not so, as a quick test with a garden hose will prove. Applying internal pressure, however great, will not enable the hose to stand on end like a charmed snake! It will not even straighten it out on the ground. (The slight tendency for the hose to rearrange its position as pressure is applied is the result of a large secondary effect; the pressure causes a slight increase in diameter with corresponding increase in bending stiffness.) Pressure end load has no influence on the required riser top tension.

Buckling of pipes with expansion joints

Figure 3–6 shows a tube with an expansion joint, subjected to internal pressure. Some have difficulty believing that such a tube can ever buckle under the influence of internal pressure, particularly with the expansion joint as shown in figure 3–6, since the force in the tube wall T_{tw} is zero.

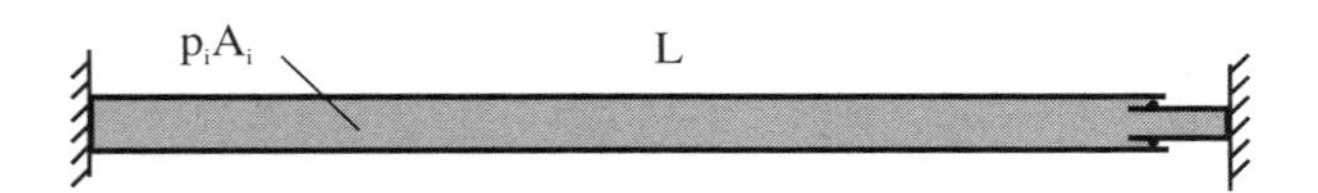

Fig. 3–6. *Tube under pressure with sliding expansion joint*

The tube will buckle when the effective axial compressive load P_e reaches the Euler buckling load ($P_E = \pi^2(EI)/L^2$). However, the effective axial compressive load P_e is equal to the total axial force in the tube wall plus internal fluid column. This is equal to $p_i A_i$, since the force in the tube wall is zero. Hence, buckling will occur for

$$p_i A_i = \frac{\pi^2}{L^2} EI \qquad (3.3)$$

The way in which an internal pressure can apply a destabilizing lateral load to a tube has been explained already (see "Lateral loads resulting from axial forces in fluid columns"). More prosaically, the tendency to buckle can also be explained by considering the tube *internal volume*. An increase in pressure will have the effect of trying to force more fluid into the tube.

The only way the tube can increase in volume is by buckling, which allows the tube end to slide partially out of its housing at the expansion joint.

Forces in connectors

Figure 3–7*a* shows schematically a bolted connector, and figure 3–7*b* shows a lugged connector of the breechblock type in the closed position with the lugs engaged. The problem is to calculate the forces in the bolts and lugs, which in both cases depend on the dimensions of the *connector seal.*

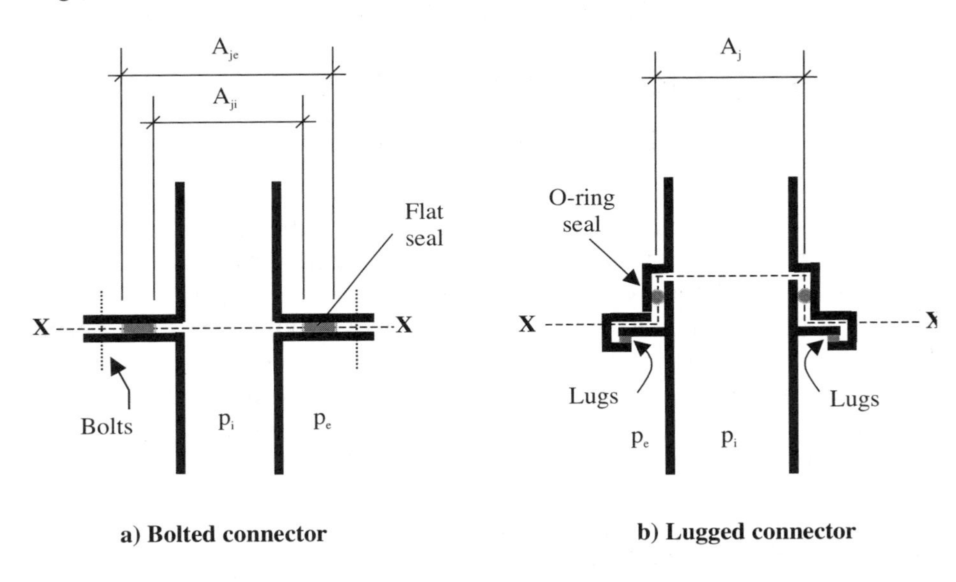

Fig. 3–7. *Forces in connector bolts and lugs*

The effective tension at the connector level can first be found by considering the top tension and the apparent weight of the intermediate riser segment. Then, the effective tension equation, applied to a section passing through the connector seal (*X–X*), allows the force in the bolts or lugs to be found. In the case of the bolted connector of figure 3–7*a,* equation (2.19) becomes

$$T_e = \sum T_{bolts} - F_{seal} - p_i A_{ji} + p_e A_{je} \qquad (3.4)$$

where F_{seal} is the compressive force acting on the connector seal; A_{ji} and A_{je} are the internal and external areas of the seal.

In the case of the lugged connector of figure 3–7*b* equipped with an O-ring, the internal and external areas of the seal are equal, and the force on the seal is negligible. Hence,

$$T_e = \sum T_{lugs} - p_i A_j + p_e A_j \qquad (3.5)$$

where A_j is the seal area ($A_j = A_{ji} = A_{je}$).

Multi-tube Risers: Components of Effective Tension and Apparent Weight

When risers are composed of more than one tube, their analysis can sometimes be greatly simplified by formulating the effective tension and the apparent weight of the riser as the sum of the *component effective tensions* and the *component apparent weights* of the individual tubes. This opens the way to straightforward treatment of complicated cases of multi-barrier risers with changes of internal fluids and other parameters; such cases are examined in detail in chapter 5. Two cases need to be considered here: risers composed of *separate tubes* and risers composed of *tubes within tubes*.

Risers composed of separate tubes

Equation (2.9)—and the physical interpretation in italics immediately following it—has already expressed the way to formulate the effective tension in a riser composed of more than one tube. There is plainly no difficulty in rewriting equation (2.9) as follows:

$$T_e = \sum \left(T_{tw} - p_i A_i + p_e A_e \right) \qquad (3.6)$$

The effective tension in the riser is then the sum of the effective tensions in the individual tubes. Likewise, from equation (2.10), the apparent weight of the riser is the sum of the apparent weights of the individual tubes:

$$w_a = \sum \left(w_t + w_i - w_e \right) \qquad (3.7)$$

Equations (3.6) and (3.7) are straightforward and have been found useful by designers of drilling risers.[3] They allow quick verification of the way that pressure changes in one tube, such as a kill and choke line, may influence the wall tension in the main tube.

Risers composed of tubes within tubes

The decomposition of effective tension and apparent weight for risers consisting of *tubes within tubes* is more difficult than for separate tubes, but it is still possible. While it may appear complicated, such decomposition can greatly simplify the analysis of certain complex riser geometries. For example, it simplifies the analysis of risers composed of tubes within tubes, made of different materials, locked together at their extremities, and subject to changes of pressure, temperature, and internal fluids.

Figure 3–8 shows a section of a riser tube within tube. The outer tube (1) has internal and external cross-sectional areas A_{i1} and A_{e1}, while the inner tube (2) has internal and external cross-sectional areas A_{i2} and A_{e2}. As shown in figure 3–8, p_i, p_a, and p_e are respectively the internal pressure in tube 2, the annular pressure, and the external pressure outside tube 1. Note that the cross-sectional area of the annulus is equal to $(A_{i1} - A_{e2})$.

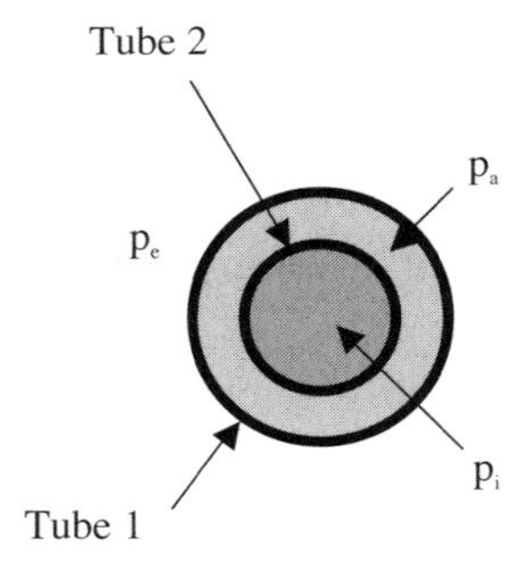

Fig. 3–8. *Riser consisting of tube within tube*

Equation (2.9) can be applied to give the effective tension in the combined system, where T_{tw1} and T_{tw2} are the true wall tensions in tubes 1 and 2, respectively:

$$T_e = T_{tw1} + T_{tw2} + (-p_i A_{i2}) + [-p_a (A_{i1} - A_{e2})] - (-p_e A_{e1}) \qquad (3.8)$$

Equation (3.8) can be rewritten as follows:

$$T_e = (T_{tw1} - p_a A_{i1} + p_e A_{e1}) + (T_{tw2} - p_i A_{i2} + p_a A_{e2}) \tag{3.9}$$

The first parenthesized expression on the right-hand side of equation (3.9) contains terms that relate only to tube 1, and the second parenthesized expression contains terms that relate only to tube 2. Hence, the two parenthesized expressions in equation (3.9) give the effective tension components T_{e1} and T_{e2} of the two tubes, which can be calculated separately. Noting that the annular pressure p_a is the internal pressure for tube 1 and the external pressure for tube 2, the effective tension component of each tube can be written as follows, where the external subscript "n" is the tube number:

$$T_{en} = (T_{tw} - p_i A_i + p_e A_e)_n \tag{3.10}$$

The effective tension for a riser bundle composed of any number of tubes, which may or may not be within each other, can be found by summing all the components expressed by equation (3.10). When calculating the effective tension components, it is important to note that:

- The internal area A_i of each tube must be taken as the full internal area, regardless of whether it has other tubes within it.
- When calculating the axial force in the displaced fluid column ($p_e A_e$), the pressure in the fluid in which the tube is really immersed must be used (external fluid for tube 1; annular fluid for tube 2).

The riser apparent weight w_a also needs careful attention. From equation (2.10),

$$w_a = w_{t1} + w_{t2} + w_{i2} + w_{annulus} - w_{e1} \tag{3.11}$$

where w_{t1}, w_{t2}, w_{i2}, $w_{annulus}$, and w_{e1} are the weight per unit length of tube 1, tube 2, tube 2 internal fluid, the annular fluid between the tubes, and the external fluid displaced by tube 1, respectively.

The weight per unit length of the annular fluid can be written as $w_{annulus} = w_{i1} - w_{e2}$, where w_{i1} is the weight per unit length of annular fluid that would entirely fill tube 1 and w_{e2} is the weight per unit length of annular fluid displaced by tube 2. Thus, equation (3.11) can be rewritten as

$$w_a = (w_{t1} + w_{i1} - w_{e1}) + (w_{t2} + w_{i2} - w_{e2}) \tag{3.12}$$

The first parenthesized expression on the right-hand side of equation (3.12) contains terms that relate only to tube 1, and the second parenthesized expression contains terms that relate only to tube 2. Hence, the two parenthesized expressions of equation (3.12) give the apparent weight components (w_{a1} and w_{a2}) of the two tubes, which can be calculated separately.

Apparent weight w_{i1} is the weight per unit length of internal fluid that would entirely fill tube 1, and w_{e2} is the weight per unit length of external fluid displaced (in the annulus) by tube 2. The apparent weight component w_{an} of each tube can be written as follows, where the external subscript "*n*" is the tube number:

$$w_{an} = \left(w_t + w_i - w_e\right)_n \qquad (3.13)$$

In formulating the components of apparent weight for each tube according to equation (3.13) it is important to note:

- w_i is the weight per unit length of internal fluid that would *entirely fill the tube cross-section* (ignoring the possible presence of other tubes within it)
- When calculating the weight per unit length of fluid displaced w_e, the density of the fluid *in which the tube is really immersed* must be used.

If the bullet points in this section are carefully respected, the effective tension and the apparent weight of a multi-tube riser can be decomposed into the components from each tube—no matter the number of tubes and no matter whether the riser consists of separate tubes or tubes within tubes or a combination of the two.

Effective Tension and Riser Dynamics

In chapter 2 (preceding equation [2.9]), it was stressed that the only assumption used in deriving the effective tension equation was that the three systems shown in figure 2–6 (tube, internal fluid column, and displaced fluid column) should be in equilibrium. The reader has the right, therefore, to question whether effective tension is valid for dynamic cases. In fact, effective tension is still valid, providing that dynamic loads are considered to be superimposed on static loads. Theoretically, several dynamic loads need to be considered.

Influence of internal flow

If the internal fluid is flowing uniformly, the rate of change of momentum ($\dot{m}_i u_i$) entering and leaving the riser should be included in the effective tension equation, just as for the pump/turbine example alluded to in figure 2–8. This is well specified in the Det Norske Veritas (DNV) code, where the change of momentum $\dot{m}_i u_i$ is given as $\rho_i A_i u_i^2$.[4] The effective tension equation then becomes

$$T_e = T_{tw} - p_i A_i + p_e A_e - \rho_i A_i u_i^2 \qquad (3.14)$$

where ρ_i, A_i, and u_i are the internal fluid mass density, the internal cross-sectional area, and the internal fluid velocity, respectively.

As the DNV points out, this additional term does not change the effective tension T_e, which can still be calculated from top tension and apparent weight. However, it does modify T_{tw}, even though this modification may be small. Frictional forces between the fluid and the pipe wall should also theoretically be included, although most codes ignore them.

Likewise, if the internal fluid is flowing and the riser is curved, radial (centrifugal) forces will be generated plus Coriolis forces if the riser itself is moving. However, according to the DNV, model tests (conducted by SINTEF) of a U-shaped flexible hose have shown that steady flow has negligible influence on riser dynamics.[5]

Hydrodynamic forces

Hydrodynamics is a highly specialized subject that is treated in detail in many works in the literature.[6] The subject can be given only very summary treatment here. Hydrodynamic loads on small-diameter submerged objects such as risers have been almost exclusively calculated using the *Morison equation* for more than 50 years, even though the equation has been considered controversial for many years. Appendix C presents the Morison equation in its initial form (for stationary objects) and shows how the version for nonstationary objects such as risers can be derived.

Basically, if the riser itself is moving laterally with velocity (v) and acceleration ($\dot{v}$), in a fluid stream that itself is moving with velocity (u) and acceleration ($\dot{u}$), then Morison's equation for the hydrodynamic force per unit length acting on the riser can be written in two ways:

$$f(x) = \frac{1}{2}\rho C_D \phi (u - v)|u - v| + \rho A_e \dot{u} + (C_M - 1)\rho A_e (\dot{u} - \dot{v}) \quad (3.15)$$

or

$$f(x) = \frac{1}{2}\rho C_D \phi (u - v)|u - v| + C_M \rho A_e \dot{u} - (C_M - 1)\rho A_e \dot{v} \quad (3.16)$$

where ρ is the fluid (mass) density, C_D is the drag coefficient, ϕ is the riser diameter, C_M is the inertia coefficient, and A_e is the riser external cross-sectional area.

The first right-hand term of equations (3.15) and (3.16) represents the drag force, and the last two make up the inertia force. Codes of practice specify one or the other of the two versions. The term $(C_M - 1)\ \rho A_e$ is frequently termed the *added mass* for convenience, since it has the units of mass and the same acceleration as the riser itself. In the literature, $(C_M - 1)$ is often given the symbol C_m and is called the *added-mass coefficient*. The value of C_M is typically close to 2. Hence, C_m is typically close to 1.

The force on an element of a riser as given by the Morison equation can be considered to be the resultant of three dynamic pressure fields, which have to be superimposed on the static pressure field. The four pressure fields can then be summarized as follows:

- The static pressure field, studied by Archimedes
- The dynamic pressure field in the fluid in the absence of the riser (the middle term of equation [3.15])
- The pressure field resulting from the presence of the riser and the *relative acceleration* of the flow with respect to it (the last term of equation [3.15])
- The pressure field resulting from the disturbed flow relative to the riser, treated as if it were of *constant velocity* (the first term of equation [3.15])

As can be seen from the equations, the forces from the four fields are in effect calculated separately, and their resultants are summed. It is remarkable that such a procedure leads to acceptable results.

Principle Reasons for Confusion about Effective Tension

When confusion persists about a subject. it is often helpful to list the reasons for the confusion. Readers probably have their own ideas about the reasons for persistent confusion about effective tension, but the principal reasons as I perceive them are the following:

- *The double convention.* The fact that both tension and pressure are treated as positive by convention must be the principal source of confusion. Notably, there is no confusion about the parallel concept of apparent weight, to which the double convention does not apply.
- *End loads and end effects.* Trying to interpret effective tension in terms of end loads and end effects appears to confuse, rather than clarify, the issue for all but the simplest cases.
- *Buoyancy.* This is a concept originally derived by Archimedes that gives the global force on a submerged object, resulting from a closed pressure field. Trying to apply the concept to unclosed pressure fields acting on riser segments can lead to great confusion. The notion of buoyancy is already integrated into the concepts of effective tension and apparent weight, and requires no separate consideration.
- *Terminology.* Some confusion results from the different terminology used for riser tension and weight. This book defines and uses the terms effective tension, true wall tension, and apparent weight, which are commonly used in the literature today. However, other terms are sometimes used for these concepts, which can lead to confusion. True wall tension is sometimes referred to as true tension, real tension, actual tension, or absolute tension. Effective tension is sometimes called fictive tension, or absolute tension less end cap load. Apparent weight is sometimes called effective weight or wet weight. It has even been called the effective weight of the effective mass!
- *Known lateral stresses between the pipe and the internal fluid column.* Because the lateral stresses (pressures) between the pipe wall and the internal fluid are known, it is legitimate to define the pipe as the system and treat the lateral pressures as external loads. This

can obscure the fact that it is actually easier and clearer to treat the pipe plus internal fluid column as part of a combined system. In the analysis of analogous steel/concrete column problems (as mentioned in chap. 2), civil engineers adopt this approach naturally and arrive at their conclusions with less confusion. They do so because they have little choice! They do not know the interface stresses between the steel and concrete precisely. Their lack of knowledge actually forces them to adopt a simpler approach to their very similar problem.

Summary

Many difficulties associated with effective tension have been discussed. These include the application and the relevance of end loads and end effects to general pipe and riser cases; the confusion that can result from trying to adapt the concept of buoyancy to continuous structures such as risers; the application of effective tension to various problems; and a number of recurring questions.

Ways have been presented of decomposing the effective tension of multi-tube risers into the component contributions of the individual tubes. This opens the way to the straightforward analysis of multi-barrier risers. The application of effective tension to dynamic cases has also been discussed.

Further difficulties occur when the effective tension is influenced by axial strain, either because the latter is prevented or because it modifies the tension. Such cases are treated in chapter 5. First, though, pipe wall stresses need to be discussed. This is the subject of the next chapter.

References

[1]Sparks, C. P. 1984. The influence of tension, pressure and weight on pipe and riser deformations and stresses. *Transactions of ASME. Journal of Energy Resources Technology*. 106 (March), 46–54.

[2]Do, C. 1983. Stabilité verticale d'une poutre dans un champs de pression hydrostatique. *Journal de Mécanique théorique et appliquée.* 2 (1), 101–111.

[3]Gaillard, C., J. Guesnon, C. Sparks, and Y. Kerbart. 2000. Riser technology for ultra deep water drilling. Paper presented at the Deep Offshore Technology Conference, New Orleans.

[4]Offshore Standard DNV-OS-F201. 2001. *Dynamic Risers.*

[5]Ibid.; SINTEF. 2006. *FPS 2000 Flexible Pipes and Risers: Handbook on Design and Operation of Flexible Pipe*s. SINTEF report STF70 A92006.

[6]Sarapkaya, T. 1981. *Mechanics of Wave Forces on Offshore Structures.* New York: Van Norstrand;
Chakrabarti, S. K. 1987. *Hydrodynamics of Offshore Structures.* Computational Mechanics Publications.WIT Press, Southampton, UK;
Molin, B. 2002. *Hydrodynamique des Structures Offshore*. Editions Technip, Paris, France;
Faltinsen, O. M. 1990. *Sea Loads on Ships and Offshore Structures.* Cambridge University Press, Cambridge, UK.

4

Pipe and Riser Stresses

This chapter is devoted to the stresses induced by a combination of tension and internal and external pressures in *circular cylindrical* pipes and risers made from elastic materials. Whereas, the axial and circumferential *forces* induced by tension and pressure in the wall of a pipe or riser are independent of the pipe material properties, the distribution of *stresses* across the section does depend on the material. In the case of flexible pipes or fiber-reinforced composite pipes, the distribution of stresses will depend on the details of the pipe wall construction; hence, the distribution can be controlled to a certain extent by the designer. For pipes made from isotropic elastic materials, the distribution of stresses depends on mechanical principles over which the designer has no control. The governing equations were first formulated by Lamé in 1831.

Stresses in Thick-Walled Elastic Pipes

Lamé applied differential calculus to the analysis of the internal stresses and strains induced in a thick-walled isotropic elastic pipe by axial tension and internal and external pressures, while working on artillery problems for the czar of Russia. The analytical formulae that he derived can be found in many text books and are also given in appendix D.[1] He found that both the circumferential stress σ_c and the radial stress σ_r vary as a function of the radial distance from the tube axis, but their sum is constant at all points of the cross-section (see equation [D.35]). Using symbols p_i and p_e for the internal and external pressures and A_i and A_e for the internal and external cross-sectional areas, he found

$$\frac{\sigma_c + \sigma_r}{2} = \frac{p_i A_i - p_e A_e}{A_e - A_i} \tag{4.1}$$

Appendix D shows that the integral of the mean of the circumferential plus radial stress across the total section of a circular cylindrical pipe is *always* equal to $p_i A_i - p_e A_e$, whatever the pipe material (see equation [D.6]). In contrast, equation (4.1) applies to *all points* of the section, but only for pipes made of isotropic elastic materials.

The fact that $(\sigma_c + \sigma_r)/2$ is constant at all points of the section of isotropic elastic pipes implies that there is no tendency for the cross-section to warp owing to the Poisson effect. Consequently, for such pipes, the axial stress is also constant across the section (see equation [D.37] and the following discussion).

The axial tension T_{tw} in the pipe wall can be decomposed into two parts—namely $(p_i A_i - p_e A_e)$, which is the axial wall tension that would be induced by the pressures if the pipe were locally capped (cf. tube *e* of fig. 3–2), and T_e, which is the effective tension (cf. equations [2.19] and [3.2]):

$$T_{tw} = (p_i A_i - p_e A_e) + T_e \tag{4.2}$$

Dividing the axial tension T_{tw} by the wall section gives the axial stress:

$$\sigma_{tw} = \frac{T_{tw}}{A_e - A_i} \tag{4.3}$$

With substitution for T_{tw} from equation (4.2), the axial stress σ_{tw} can be decomposed into the two components given in equation (4.4) and can be written alongside Lamé's equations for the circumferential and radial stresses σ_c and σ_r, respectively given in equation (4.5) and (4.6):

$$\sigma_{tw} = \sigma_p + \sigma_{le} \tag{4.4}$$

$$\sigma_c = \sigma_p + \tau \tag{4.5}$$

$$\sigma_r = \sigma_p - \tau \tag{4.6}$$

In equations (4.4–4.6), σ_{le} is the effective stress, σ_p is the end effect stress, and τ is the in-wall shear stress:

$$\sigma_{le} = \frac{T_e}{A_e - A_i} \tag{4.7}$$

$$\sigma_p = \frac{p_i A_i - p_e A_e}{A_e - A_i} \tag{4.8}$$

$$\tau = \frac{(p_i - p_e) A_i A_e}{(A_e - A_i) A_r} \tag{4.9}$$

As before, p_i, p_e are the internal and external pressures; $A_i = \pi r_i^2$, $A_e = \pi r_e^2$, and $A_r = \pi r^2$, where r_i, r_e are the internal and external radii and r is the radius at the point in the pipe wall under consideration.

Note from equations (4.5) and (4.6) that $\sigma_c + \sigma_r = 2\sigma_p$. Since σ_r is equal to $-p_i$ and $-p_e$ on the inner and outer surfaces, respectively, the circumferential stresses σ_{ci} and σ_{ce} at those surfaces are given *exactly* by equations (4.10) and (4.11):

$$\sigma_{ci} = 2\sigma_p + p_i \tag{4.10}$$

$$\sigma_{ce} = 2\sigma_p + p_e \tag{4.11}$$

Of the three stresses (σ_{le}, σ_c, τ) given by equations (4.7–4.9), only τ varies across the pipe wall section, having its greatest value at the internal surface of the tube. The *end effect stress* σ_p is common to all three principal stresses given by equations (4.4–4.6).

Figure 4–1 shows the Mohr's circle of stress, corresponding to equations (4.5) and (4.6). The radial and circumferential stresses (σ_r and σ_c, respectively) are principal stresses, and τ is the radius of the stress circle. Therefore, τ is the maximum shear stress, acting on planes at 45° to radial directions, for which the direct stresses are both equal to the end effect stress σ_p.[2] Stress circles, for all points of the section, have the same center (σ_p, 0). Their radius τ varies with $1/r^2$. Hence, τ always has its greatest value at the inner surface of the pipe.

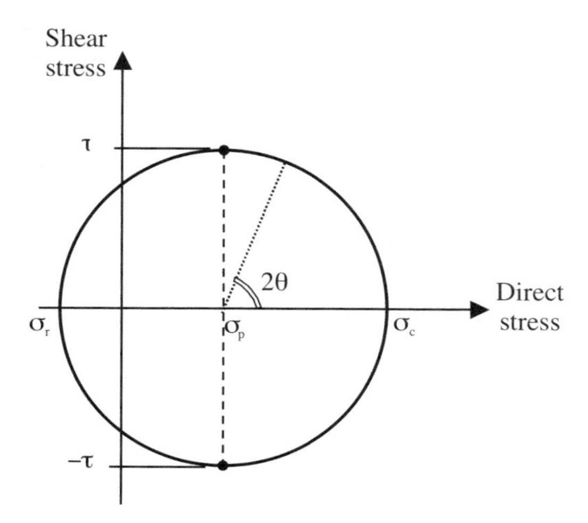

Fig. 4–1. *Mohr's circle of stress*

Thus, the triaxial stresses of equations (4.4–4.6) are precisely equivalent to a system composed of three pure stresses: a pure *hydrostatic* stress σ_p, a pure *shear* stress τ, and a pure *axial* stress σ_{le}. The two equivalent stress systems are shown graphically in figure 4–2.

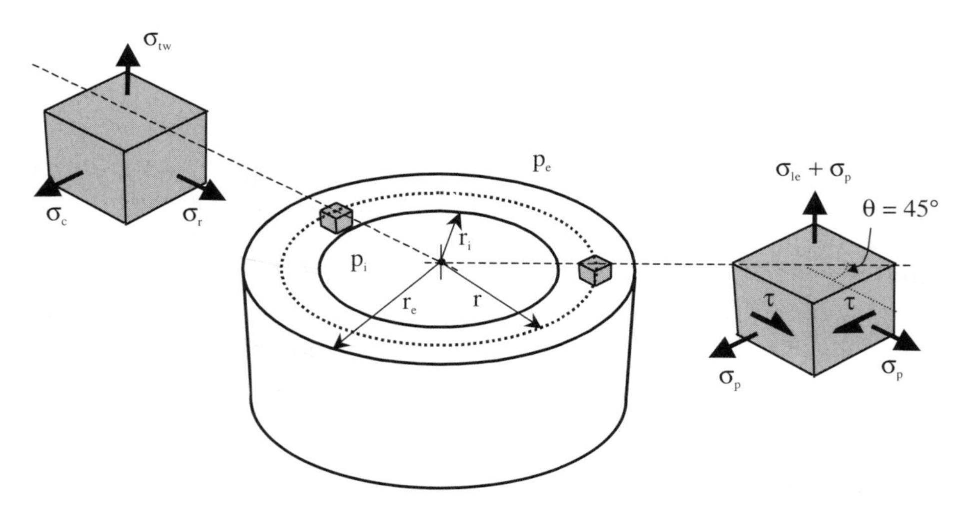

Fig. 4–2. *In-wall stresses—two equivalent stress systems*

Since σ_{le} is equal to the effective tension divided by the pipe wall section, it is called the *effective stress*. This name should not mislead the reader into thinking that it is the only stress in the pipe wall that needs to be considered.

The effective stress σ_{le}, as defined in equation (4.7), is shown on the right-hand stress cube of figure 4–2. It is the amount by which the axial wall stress σ_{tw} exceeds the in-wall hydrostatic stress σ_p. Such a stress is a *deviator stress,* which is a concept familiar to engineers working in other fields. For example, an important parameter in a soil mechanics triaxial test is the difference between the vertical and the horizontal stresses acting on the soil sample, which is also a deviator stress.[3]

Effective Stress and Excess Stress

Effective tension T_e was seen to be the sum of the axial forces in the pipe plus internal fluid column, less the axial force in the displaced fluid column (see the italicized text following equation (2.9)). The reader may therefore be surprised to find effective stress σ_{le}, which is equal to the effective tension divided by the wall section $T_e/(A_e - A_i)$, appearing as a component of the axial stress *in the pipe wall.*[4]

It follows from the equivalence between the effective tension and the excess tension mentioned in chapter 3, preceding equation (3.2). Figure 4–3 shows this equivalence graphically. External pressure has been excluded for clarity. Figure 4–3*a* shows the components of the effective tension, as given in italics following equation (2.9).

The left part of Figure 4–3*b* shows the decomposition of the forces in the pipe wall as given by equation (4.2). The *effective tension* T_e of figure 4–3*a* is plainly always equal to the *excess tension* T_e shown in the left-hand part of figure 4–3*b*. This is no surprise since the equivalence between effective tension and excess tension has already been noted in chapter 3 preceding equation (3.2).

The right-hand part of figure 4–3*b* shows the decomposition of axial wall stress. The stress σ_{le} is equal to T_e/A. It can therefore be considered to be the excess stress (i.e., in excess of the pressure end effect stress σ_p) or the effective stress since it is equal to the effective tension T_e divided by the wall section. The name *effective stress* is preferred since it immediately links σ_{le}

to the effective tension. Note that the names *neutral axial stress* (for σ_p) and the *excess stress above the neutral value* (for σ_{le}) have sometimes been used elsewhere.[5]

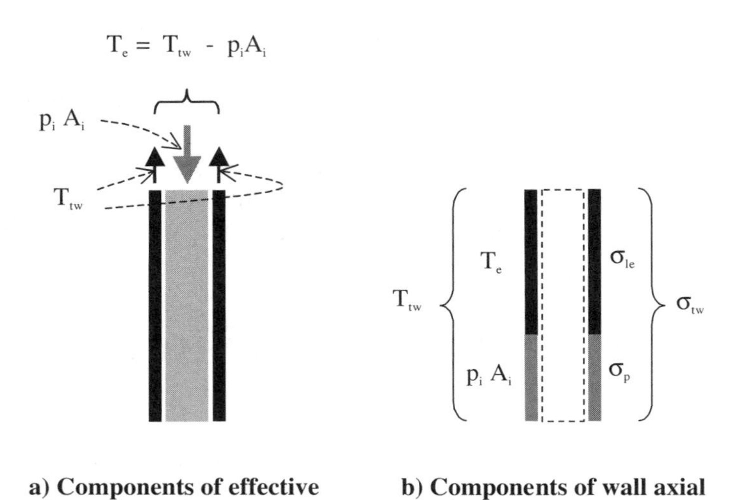

a) Components of effective tension

b) Components of wall axial forces and stresses

Fig. 4–3. *Components of axial forces and stresses in a pipe under internal pressure*

Von Mises' Equivalent Stress

The combination of stresses that provokes the onset of yield in ductile materials was the subject of research and debate for many years. Tresca's maximum stress criteria, formulated in 1876, was used for many years.[6] Today the maximum distortional energy criterion, originally formulated by von Mises in 1913, is considered to be the most accurate criterion for ductile materials.[7] This criterion is represented by the von Mises' equivalent stress σ_{vm}, which for the general cases of triaxial stresses is given by

$$2\sigma_{vm}^2 = (\sigma_1 - \sigma_2)^2 + (\sigma_2 - \sigma_3)^2 + (\sigma_3 - \sigma_1)^2 + 6(\tau_{12}^2 + \tau_{23}^2 + \tau_{31}^2) \quad (4.12)$$

Equation (4.12) can be applied to the stress cubes of figure 4–2. The left-hand cube shows the principal stresses, for which shear stresses are zero, giving

$$2\sigma_{vm}^2 = (\sigma_{tw} - \sigma_c)^2 + (\sigma_c - \sigma_r)^2 + (\sigma_r - \sigma_{tw})^2 \qquad (4.13)$$

Alternatively, the von Mises' stress can be formulated by considering the right-hand cube of figure 4–2. Since σ_p is a hydrostatic stress, it produces *no distortion* and therefore does not appear in the equation. The only stresses that contribute to the von Mises' stress are the pure axial stress σ_{le} and the pure shear stress τ. Hence, the von Mises' stress is given by equation 4.14, which can also be obtained by substituting equation (4.4–4.6) into equation (4.13):

$$\sigma_{vm}^2 = \sigma_{le}^2 + 3\tau^2 \qquad (4.14)$$

The stresses σ_{tw} and σ_{le} in equations (4.13) and (4.14) are *mean axial stresses,* respectively equal to the true wall tension and the effective tension divided by the wall section. In reality, a riser is generally also subject to moments that induce bending stresses σ_{ab}. These have to be added to the axial stresses when verifying the von Mises' stress, which then has to be checked at both the inner surface, where the shear stress τ is maximum, and the outer surface, where the bending stress σ_{ab} is maximum.

Hence, when a riser is subject to bending, equations (4.13) and (4.14) have to be replaced by equations (4.15) and (4.16), respectively:

$$2\sigma_{vm}^2 = \left(\overline{\sigma_{tw} + \sigma_{ab}} - \sigma_c\right)^2 + (\sigma_c - \sigma_r)^2 + \left(\sigma_r - \overline{\sigma_{tw} + \sigma_{ab}}\right)^2 \qquad (4.15)$$

$$\sigma_{vm}^2 = (\sigma_{le} + \sigma_{ab})^2 + 3\tau^2 \qquad (4.16)$$

When equation (4.13) or (4.15) is used, care has to be taken to respect the double-sign convention, which implies that *positive* pressures (internal and external) induce *negative* radial stresses.

Position of Codes of Practice with Respect to Stresses

Von Mises' equivalent stress

Most codes require the von Mises' equivalent stress to be checked. It is generally specified in accordance with equation (4.12) or (4.13), with the mention that bending stresses should be included in the axial stresses and that torsion and shear stresses are assumed to be negligible.

Some codes quote equation (4.16), without explaining the equivalence of the stress systems shown in figure 4–2. They express σ_{le} as the effective tension divided by the wall section and simply state that "a little algebra" leads to that equation.

Principal stresses

Codes of practice are concerned first with safety. As a result, they tend to break down stresses into components to which they apply different factors according to the associated risk. Hence, they tend not to use the exact Lamé stress formulae for thick-walled pipes. They decompose the axial and circumferential stresses into two components: a *membrane stress,* which is the mean stress in the section, and a *bending stress,* which accounts for the variation of the stress across the section.

For the case of axial stresses, the membrane stress is simply the mean stress, or the axial tension divided by the wall section. The axial bending stress is the classical bending stress, equal to the moment times the distance r from the neutral axis, divided by the second moment of area of the section (Mr/I). It is also equal to the Young's modulus times the distance from the neutral axis, times the curvature (Er/R).

For circumferential stresses, the membrane stress σ_{cm} is equal to the circumferential force divided by the wall thickness:

$$\sigma_{cm} = \frac{p_i r_i - p_e r_e}{r_e - r_i} \qquad (4.17)$$

Note that the circumferential membrane stress σ_{cm} is *not equal* to the mean of the circumferential stresses σ_c at the inner and outer surfaces given by equation (4.5), since σ_c does not vary linearly across the section.

The sum of the membrane stress σ_{cm} and the circumferential bending stress σ_{cb} is equal to the circumferential stress given by equation (4.5). Hence, at the inner surface, $\sigma_{cm} + \sigma_{cbi} = \sigma_p + \tau_i$, and at the outer surface, $\sigma_{cm} + \sigma_{cbi} = \sigma_p + \tau_e$. Note that the circumferential bending stress is positive (tensile) at the inner surface and negative (compressive) at the outer surface. From equations (4.8), (4.9), and (4.17), the circumferential bending stresses simplify to

$$\sigma_{cbi} = \frac{r_e}{r_e + r_i}(p_i - p_e) \qquad (4.18)$$

at the inner surface and

$$\sigma_{cbe} = -\frac{r_i}{r_e + r_i}(p_i - p_e) \qquad (4.19)$$

at the outer surface.

Plasticity allows some redistribution of bending stresses, which is not the case for membrane stresses. Hence, the codes require higher safety factors for membrane stresses than for bending stresses.

Two Particular Yield Problems

Yield of tubes under pressure with and without end effect

Pressure end effect has been discussed in detail in the first section of chapter 3. Figure 4–4 shows two identical tubes under internal pressure p_i. Tube *a* is capped and is subject to pressure end effect, whereas tube *b* is uncapped. The question is which tube will reach yield first, according to the von Mises' yield criterion, as pressure is increased.

Fig. 4–4. *Capped and uncapped tubes under pressure*

In fact, it is tube *b,* without end effect, that reaches the yield criterion first. This may seem surprising at first, since the true wall axial stress is zero and therefore the tube appears to be subjected to less stress than tube *a*. However, the result is obvious from the stress system of the right-hand cube of figure 4–2. The effective tension in tube *a* is zero. Therefore, the effective stress σ_{le} is also zero. Hence, for tube *a,* from equation (4.14),

$$\sigma_{vm}^2 = 3\tau^2 \quad (4.20)$$

For tube *b,* the effective tension is *not zero*. The true wall tension is zero; hence, the true wall stress σ_{tw} is also zero. Therefore, for tube *b,* from equations (4.4) and (4.14),

$$\sigma_{le} = -\sigma_p \quad (4.21)$$

and

$$\sigma_{vm}^2 = \sigma_p^2 + 3\tau^2 \quad (4.22)$$

The in-wall shear stress τ given by equation (4.9) is the same for both tubes. Hence, by comparison of equations (4.20) and (4.22), it can be seen that the von Mises' stress is greater for tube *b,* without end effect, than for tube *a,* with end effect. Thus, tube *b* yields before tube *a*.

Yield of tubes under pressure with axial load

Figure 4–5 shows two identical tubes under internal pressure p_i. One tube is subject to an axial *tension* F_e. The other is subject to an axial *compression* F_e of the same magnitude. The question is which tube will yield first, according to the von Mises' yield criterion, as the pressure is increased.

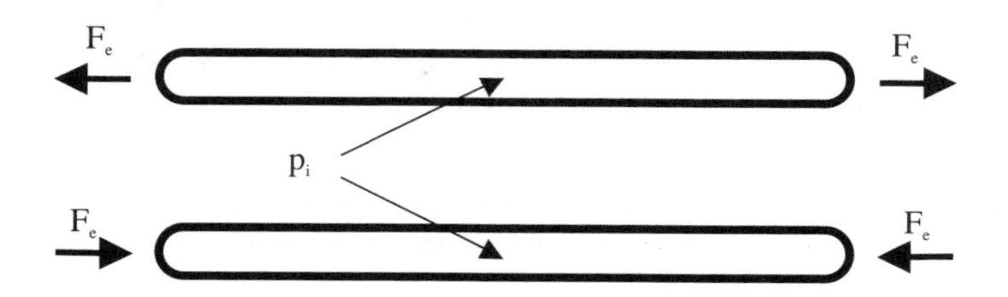

Fig. 4–5. *Identical pressurized tubes under axial tension and compression*

Both tubes are subject to the same circumferential and radial stresses, but the *true wall axial stress* is plainly less in the lower tube, under external axial compression, than in the upper tube, under tension. Nevertheless, the von Mises' equivalent stresses are the *same* for the two tubes.

The preceding is immediately obvious if the von Mises equivalent stress is calculated using equation (4.14). The in-wall shear stress τ is the same for both tubes. For the two tubes, the effective tensions—and hence the effective stresses σ_{le}—differ only in sign. Therefore, the effective stresses are given by

$$\sigma_{le} = \frac{F_e}{A_e - A_i} \qquad (4.23)$$

for the upper tube and

$$\sigma_{le} = -\frac{F_e}{A_e - A_i} \qquad (4.24)$$

for the lower tube. Since σ_{le}^2 is the same for both tubes, the von Mises' equivalent stress given by equation (4.14) is also the same for both tubes. The two tubes reach yield for the same values of internal pressure.

Numerical Example and Use of Excel File *Riser-Stresses.xls*

The following is a numerical example of a stress calculation for a particular riser (22 in. outer diameter [OD], 1 in. wall) subject to tension, pressures, and bending moment. The effective tension T_e is *specified,* and the true wall tension T_{tw} is then deduced from the pressures and cross-sectional areas by using equation (2.19), as explained at the end of chapter 2.

Example data:

Effective tension T_e = 1,750 kN

Moment M = 200 kN-m

Internal radius r_i = 0.2413 m; external radius r_e = 0.2667 m

Internal pressure p_i = 35,000 kPa; external pressure p_e = 10,500 kPa

Riser resulting characteristics:

Internal area $A_i = 0.18292\ m^2$; external area $A_e = 0.22346\ m^2$

Cross-sectional area $A = A_e - A_i = 0.04054\ m^2$

Second moment of area $I = 0.001311\ m^4$

Axial force and stresses:

Fluid column forces: internal, $p_i A_i = 6{,}402$ kN;
displaced, $p_e A_e = 2{,}346$ kN

True wall tension: $T_{tw} = T_e + (p_i A_i - p_e A_e) = 5{,}806$ kN

True wall stress: $\sigma_{tw} = T_{tw}/A = 143{,}227$ kPa

Effective stress: $\sigma_{le} = T_e/A = 43{,}171$ kPa

Axial bending stresses: $\sigma_{abi} = Mr_i/I = 36{,}814$ kPa;
$\sigma_{abe} = Mr_e/I = 40{,}690$ kPa

In-wall (σ_p *and* τ) *and circumferential* (σ_c) *stresses (equations [4.5], [4.8], and [4.9]):*

$$\sigma_p = \frac{p_i A_i - p_e A_e}{A_e - A_i} = 100{,}056 \text{ kPa}$$

$$\tau_i = \frac{(p_i - p_e)A_e}{A_e - A_i} = 135{,}056 \text{ kPa}; \qquad \tau_e = \frac{(p_i - p_e)A_i}{A_e - A_i} = 110{,}556 \text{ kPa}$$

$$\sigma_{ci} = \sigma_p + \tau_i = 235{,}113 \text{ kPa}; \qquad \sigma_{ce} = \sigma_p + \tau_e = 210{,}613 \text{ kPa}$$

Von Mises' stress (σ_{vm}) *calculated using principal stresses (equation [4.15]):*

	Internal	**External**
$\sigma_{tw} + \sigma_{ab}$	180,042 kPa	183,917 kPa
σ_c	235,113 kPa	210,613 kPa
σ_r	–35,000 kPa	–10,500 kPa
σ_{vm}	247,221 kPa	209,047 kPa

Von Mises' stress (σ_{vm}) *calculated using effective stress (equation [4.16]):*

	Internal	External
$\sigma_{le} + \sigma_{ab}$	79,985 kPa	83,861 kPa
τ	135,056 kPa	110,556 kPa
σ_{vm}	247,221 kPa	209,047 kPa

Circumferential stresses (from equations [4.17]–[4.19]):

	Internal	External
Membrane stress σ_{cm}	222,250 kPa	222,250 kPa
Bending stress σ_{cb}	12,863 kPa	–11,638 kPa

Check of circumferential stresses (from preceding table and equations [4.10] and [4.11]):

	Internal	External
$\sigma_{cm} + \sigma_{cb} = \sigma_c$	235,113 kPa	210,613 kPa
$\sigma_p + \tau = \sigma_c$	235,113 kPa	210,613 kPa
$2\sigma_p + p = \sigma_c$	235,113 kPa	210,613 kPa

As can be seen, equations (4.15) and (4.16) give exactly the same values for the von Mises' stress σ_{vm}. Since the von Mises' stress is of such importance and since errors are easily made, it is as well to evaluate it by both methods. Errors are particularly easy to make when using thin-wall stress formulae. The final check—consisting of a comparison of the results obtained using different ways to calculate the circumferential stresses—is a useful check against numerical errors.

All the relevant stress equations of this chapter are programmed in the Excel file *Riser-Stresses.xls,* which the reader can use to calculate riser stresses with other data. For ease of printing, a further version of the file, named *Riser-Stresses (print).xls,* is also provided.

The von Mises' stress is calculated in two ways: from the principal stresses and from the effective stress. All stress equations used are referenced in detail in the file *Riser-Stresses.xls.* The stress equations of this chapter are also used in the file *SCR-Example.xls,* which was mentioned in chapter 1. However, for the SCR application, the bottom-end pressures and effective tension are deduced from the catenary data, and the bending stresses near the extremities are deduced from the catenary sag.

Summary

Lamé's equations for the in-wall stresses in thick-walled pipes have been presented. It has been shown that the stresses in a thick-walled pipe can be represented by two equivalent systems. One involves the three principal stresses (axial, circumferential, and radial), while the other involves three superimposed pure stresses (axial, hydrostatic, and shear).

The concept of effective stress has been explained. It has been shown to be a deviator stress, which is a concept familiar to engineers in other fields such as soil mechanics. Two equally precise ways of formulating von Mises' equivalent stress have been presented, and the requirements of codes of practice have been discussed. Two particular yield problems have been clarified using effective stress. A detailed example of a pipe stress calculation has been included.

The Excel file *Riser-Stresses.xls* has been presented. It allows readers to calculate riser stresses using the equations of this chapter with their own data. The file *SCR-Example.xls* also applies the same equations to SCRs.

Pipe and riser strains and their influence on riser global behavior can now be discussed using the stress formulae of this chapter. This is the subject of the next chapter.

References

[1]Lamé, G., 1852. *Leçons sur la théorie mathématique de l'élasticité des corps.* Bibliothéque St. Geneviève, Paris;
Hill, R. 1964. *Plasticity*. Oxford University Press, Oxford, UK, 106–109;
Timoshenko, S. 1976. *Strength of Materials.* 3rd ed. Part 2. Huntington, N.Y.: Kriegas.

[2]Morrison, J. L. M., B. Crossland, and J. S. C. Parry. 1960. The strength of thick cylinders subjected to repeated internal pressure. *Journal of Engineering for Industry, ASME.* May, 143–153.

[3]Terzaghi, K., and R. B. Peck. 1948. *Soil Mechanics in Engineering Practice*. John Wiley. New York.

[4]Sparks, C. P. 1984. The influence of tension, pressure and weight on pipe and riser deformations and stresses. *Transactions of ASME. Journal of Energy Resources Technology*. 106 (March), 46–54.

[5]Lubinski, A. 1975. Influence of neutral axial stress on yield and collapse of pipe. *Journal of Engineering for Industry, Transactions of ASME.* May, 400–407.

[6]Case, J., and A. H. Chilver. 1961. *Strength of Materials.* Edward Arnold, London

[7]Burrows, W. R., R. Michel, and A. W. Ranking. 1954. A wall thickness formula for high pressure, high temperature piping. *Transactions of ASME.* April, 427–444.

5

Pipe and Riser Strains

The correct calculation of axial strains is of great importance to a number of different riser problems.[1] For near-vertical risers, axial strains influence the required stroke of riser tensioners. For risers without tensioners, differential strains between adjacent risers influence the riser profiles and, thus, possible riser interaction. Axial strains also have to be taken into account when analyzing the stability of drilling-riser kill and choke lines.

This chapter shows how the strains can be calculated for isotropic and anisotropic pipes. Some particular problems are discussed. Procedures are described for the calculation of axial strains induced by parameter changes (pressure, temperature, and internal fluid density) for single-tube risers, as well as for multi-tube risers.

Axial Strains of Thick-Walled Elastic Isotropic Pipes

For elastic isotropic pipes, the principal strains are related to the principal stresses by the standard relationships involving Young's modulus E and Poisson's ratio ν. Hence, the axial strain ε_a is given classically by

$$\varepsilon_a = \frac{1}{E}(\sigma_{tw} - \nu\sigma_c - \nu\sigma_r) \qquad (5.1)$$

where σ_{tw}, σ_c, and σ_r are the true wall axial stress, the circumferential stress, and the radial stress, respectively, as defined in the previous chapter. Since

σ_c and σ_r are not constant across the pipe wall, the mean stress values σ_c and σ_r, respectively given by equations (D.8) and (D.9) of appendix D, should be used when evaluating equation (5.1) for a thick-walled pipe.

The axial strain can be expressed precisely in terms of the end effect stress σ_p and the effective stress σ_{le}, which have been defined in the previous chapter (in equations [4.8] and [4.7], respectively). Lamé showed that the mean of the circumferential and radial stresses $(\sigma_r + \sigma_r)/2$ is constant across the section and equal to σ_p (see equation [D.35]). Hence, the axial strain ε_a can be expressed exactly:

$$\varepsilon_a = \frac{1}{E}\left(\sigma_{tw} - 2\nu\sigma_p\right) \tag{5.2}$$

The axial strain can also be expressed equally exactly in terms of effective stress σ_{le} by

$$\varepsilon_a = \frac{1}{E}\left[\sigma_{le} + (1-2\nu)\sigma_p\right] \tag{5.3}$$

The in-wall shear stress τ does not appear in equation (5.3) since it produces no axial strain. The advantage of equation (5.3) is that the axial strain induced by pressures is entirely contained in the second term, namely $\varepsilon_{a\,pressure} = (1-2\nu)\sigma_p / E$. This pressure-induced axial strain is zero if $\nu = 0.5$.

By substitution into equations (5.2) and (5.3) for σ_{tw}, σ_p, and σ_{le} (from eqq. [4.3], [4.8], and [4.7], respectively) and use of the symbol $A = A_e - A_i$, for the pipe cross-sectional area, the axial strain can be expressed in terms of the true wall tension, by equation (5.4), or the effective tension, by equation (5.5):

$$\varepsilon_a = \frac{1}{EA}\left(T_{tw} - 2\nu p_i A_i + 2\nu p_e A_e\right) \tag{5.4}$$

$$\varepsilon_a = \frac{1}{EA}\left[T_e + (1-2\nu)p_i A_i - (1-2\nu)p_e A_e\right] \tag{5.5}$$

Equation (5.5) is plainly directly applicable to the tube of figure 3–1*a,* for which T_e was seen to be the external tension, the excess tension, and the effective tension. Equation (5.5) is also applicable to the general case of a riser, subject to effective tension T_e.

Equation (5.5) decomposes the axial strain into components due to pure tension, for which the axial stiffness is equal to *EA,* and due to internal and external pressures, for which the axial stiffnesses (with respect to pressure end effects) are both equal to $EA/(1-2\nu)$. The axial strains due to internal and external pressures are entirely contained in the second and third terms, respectively, of equation (5.5). In equations (5.4) and (5.5), the effects of internal and external pressures are kept separate, for reasons that will become clear in the next section.

Axial Strains of Anisotropic Pipes

Today alternative materials such as high-performance composites are being considered for risers. Such materials are elastic but anisotropic. They have been seriously studied for various offshore applications, including risers and flow lines, since the 1980s.[2] High-performance composites pipes have exceptional mechanical properties in the form of high strength, light weight, and excellent fatigue properties. They are made from high-strength fibers set in a resin matrix. By careful orientation of the fibers with respect to the pipe axis, it is possible to obtain further interesting characteristics, such as near-zero axial strains due to pressure and thermal effects.[3]

The calculation of strains for pipes made from such materials is more complicated than for isotropic pipes, since it involves three Young's moduli and six Poisson's ratios (see equation [E.1] of appendix E).[4] The hypotheses and calculation procedures have been described in many publications.[5] For any particular design, the resulting axial stiffness under pure tension—and under pressure with end effect—should always be verified by measurements on tube samples.

Pressure-induced axial strains can be used to define *equivalent Poisson's ratios* $\overline{\nu}_i$ and $\overline{\nu}_e$ for internal and external pressure. For pipes made from anisotropic materials, the two equivalent Poisson's ratios are different. Equations (5.4) and (5.5) therefore have to be rewritten as equations (5.6) and (5.7):

$$\varepsilon_a = \frac{1}{E_a A}\left(T_{tw} - 2\overline{\nu}_i p_i A_i + 2\overline{\nu}_e p_e A_e\right) \tag{5.6}$$

$$\varepsilon_a = \frac{1}{E_a A}\left[T_e + \left(1-2\overline{\nu}_i\right)p_i A_i - \left(1-2\overline{\nu}_e\right)p_e A_e\right] \tag{5.7}$$

where E_a is the tube axial modulus, and $A = A_e - A_i$. The axial stiffness related to the internal pressure end load $p_i A_i$ then becomes $E_a A/\left(1-2\overline{\nu}_i\right)$.

Likewise, the axial stiffness related to the external pressure end load $p_e A_e$ becomes $E_a A/(1-2\overline{\nu}_i)$.

The equivalent Poisson's ratios $\overline{\nu}_i$ and $\overline{\nu}_e$ so defined are related to the Poisson's ratios ν_{ca} and ν_{ra} of the composite laminate, as shown in the first section of appendix E (see eqq. [E.8] and [E.9]). The Poisson's ratios ν_{ca} and ν_{ra} can be calculated from the basic properties of the layers that make up the laminate.[6]

Hence, the procedure to obtain specific values of $\overline{\nu}_i$ and/or $\overline{\nu}_e$ is to design the laminate to give the acceptable values of ν_{ca} and ν_{ra} according to equation (E.8) and/or equation (E.9). For a composite laminate, ν_{ca} can be controlled by carefully choosing the fibers—and their orientation—in the different layers of the laminate.[7] The value of ν_{ra} is more difficult to determine and control. However, since equations (E.8) and (E.9) show that $\overline{\nu}_i$ and $\overline{\nu}_e$ depend principally on ν_{ca}, for the purposes of an initial design, ν_{ra} can be ignored, giving $2\overline{\nu}_i = \nu_{ca}(1+r_e/r_i)$ and $2\overline{\nu}_e = \nu_{ca}(1+r_i/r_e)$.

The resultant values of $\overline{\nu}_i$ and $\overline{\nu}_e$ should be verified by tests on tube samples, as described in the next section. This is particularly important if the objective is to obtain a tube with a specific equivalent Poisson's ratio—such as a value close to $\overline{\nu}_i = 0.5$, for which changes in internal pressure induce zero change in axial stretch. Equations (E.8) and (E.9) can then be used to work back to the values of ν_{ca} and ν_{ra}, which it may not have been possible to determine very precisely. The laminate design can then be modified and fine-tuned if necessary.

Note that for an in-place riser, the external pressure is constant; thus, the precise value of $\overline{\nu}_e$ is of little consequence. However, if anisotropic materials should be proposed for applications such as *riser tubings,* then the value of $\overline{\nu}_e$ will need to be taken into account when calculating the tubing axial stresses and strains resulting from changes in riser *annular pressure*.

Determination of Equivalent Poisson's Ratios for Anisotropic Pipes

Equation (5.7) gives the axial strain induced by axial tension and internal and external pressures. The axial stiffness $E_a A$ of a sample tube can be determined by measuring the strain in a standard tension test. The internal equivalent Poisson's ratio can then be easily determined from the axial strain ε_a induced by an internal pressure test of a capped tube, since from equation (5.7),

$$1 - 2\overline{\nu_i} = \frac{E_a A}{p_i A_i} \varepsilon_a \qquad (5.8)$$

It is difficult to use the same technique to determine the external equivalent Poisson's ratio, since the test would have to be carried out in a pressure vessel. Fortunately, there is a simple alternative. The principle of virtual work can be used to show that the equivalent Poisson's ratio for external pressure is always given by equation (5.9) (see equation [E.17] of appendix E):

$$\overline{\nu_e} = -\varepsilon_{ce} / \varepsilon_a \qquad (5.9)$$

where ε_a and ε_{ce} are the respective axial and external circumferential strains induced by a *pure tension test*.

Pressure-Induced Buckling of Pipes Fixed at Both Extremities

A pipe fixed at both extremities can be caused to buckle by increasing internal pressure.[8] Internal pressure can cause buckling of drilling-riser kill and choke lines if the intermediate guides are too widely spaced or insufficiently firmly attached to the riser central tube. This has been observed in the laboratory and can be seen in figure 5–1, wherein model kill and choke lines were caused to buckle.[9] In the past, internal pressure has also caused unacceptable deformations of drilling-riser peripheral lines in the field. The same phenomenon has also caused unacceptable upheaval buckling of flow lines.[10]

The phenomenon can be understood qualitatively from figure 4–3*a* of the previous chapter. Plainly, the pipe will stretch axially as internal pressure is increased, unless it is made from some special material with $\nu_i \geq 0.5$. If that axial stretch is prevented between any two points (not necessarily the extremities), the pipe will eventually buckle between those points as internal pressure is increased. This problem involves *effective* *compression* F_e, rather than effective tension, but can be treated using the preceding equations by putting $F_e = -T_e$.

Fig. 5–1. *Pressure-induced buckling of a kill and choke line (courtesy of the Institut Français du Pétrole)*

Pre-buckling behavior

The axial strain is maintained constant ($\varepsilon_a = 0$), as shown in figure 5–2*a.* As the internal pressure p_i is increased, the axial compressive force $p_i A_i$ in the internal fluid column will increase. The axial tensile force in the pipe wall T_{tw} will also increase owing to the Poisson effect, as given by equation (5.4): $T_{tw} = 2\nu p_i A_i$. The effective compression F_e, which is the sum of the two axial forces (compression positive), will also increase and can be found from equation (5.5), since $\varepsilon_a = 0$: $F_e = (1 - 2\nu)p_i A_i$.

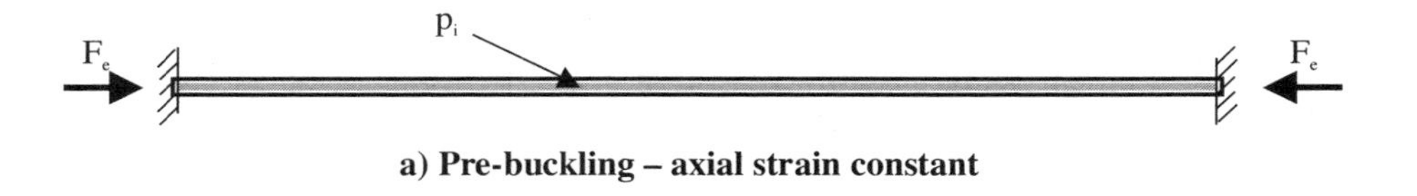

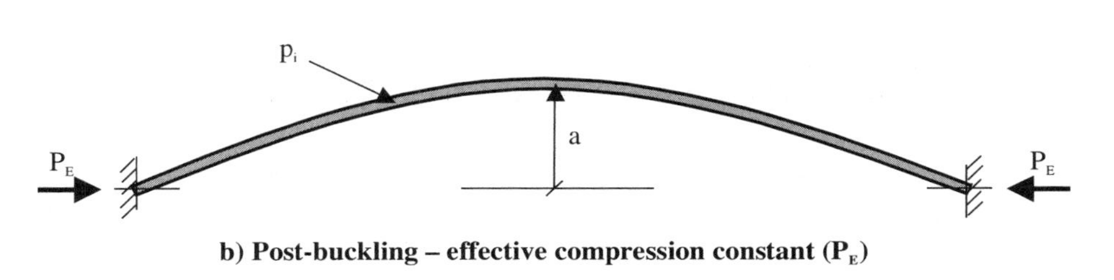

Fig. 5–2. *Pressure-induced buckling of pipes fixed at both extremities*

Buckling will occur when the effective compression in the pipe reaches the Euler buckling load: $P_E = \pi^2(EI)/L^2$. Hence, the pipe will begin to buckle when the internal pressure p_i reaches the value given by equation (5.10):

$$(1-2\nu)p_i A_i = P_E = \frac{\pi^2}{L^2}EI \tag{5.10}$$

Post-buckling behavior

As the pressure is further increased, the effective compression F_e will remain constant and equal to the Euler buckling load P_E. However, the axial strain will no longer be constant. It will increase in accordance with equation (5.5), causing the buckle to develop, as shown in figure 5–2*b*. For a sinusoidal buckle, the amplitude *a* of the pipe deflection is then related to the axial strain ε_a by

$$\varepsilon_a = \left(\pi \frac{a}{2L}\right)^2 \tag{5.11}$$

For a pipe made from anisotropic material, the preceding approach can still be used. The axial tensile force in the pipe wall T_{tw} and the effective compression F_e have to be calculated using equations (5.6) and (5.7) instead of equations (5.4) and (5.5).

For a pipe made from an anisotropic material with the special characteristic $\overline{\nu}_i = 0.5$, internal pressure can never cause the pipe to buckle, since the internal pressure–induced effective compression is always zero: $F_e = (1-2\overline{\nu}_i)p_iA_i$. For any particular internal pressure, the axial tension in the pipe wall ($T_{tw} = 2\overline{\nu}_i p_iA_i$) will then be equal to the axial compression in the internal fluid column (p_iA_i). Their sum will be zero.

Pipe Stretch Following Upending

Figure 5–3 shows a pipe of length L, before and after upending accompanied by change of internal fluid, pressure, and temperature. For the general case, the pipe stretch Δe can be calculated by integrating axial strain along its length, in the initial configuration before upending (denoted by external subscript "i") and in the final configuration after upending (denoted by external subscript "f"), and taking the difference:

$$\Delta e = e_f - e_i = \left(\int_0^L \varepsilon_a \, dx\right)_f - \left(\int_0^L \varepsilon_a \, dx\right)_i \qquad (5.12)$$

where the tension and pressure-induced axial strains are given by equation (5.4) or (5.5), to which the thermal strains have to be added.

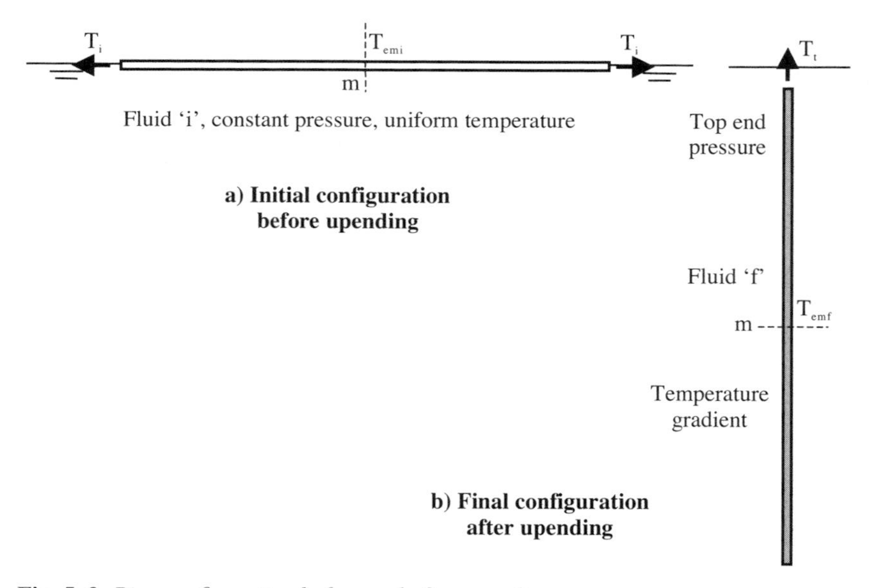

Fig. 5–3. *Pipe configuration before and after upending*

For a pipe of uniform section with fluids of constant density and constant temperature gradient (i.e., with all characteristics varying linearly over the pipe length), the problem can be greatly simplified since the axial strains can be calculated from the *mean values* of effective tension, pressure, and temperature at the midpoint of the pipe, before and after upending. From equation (5.5), the axial stretch e is given as follows:

$$e = \int_0^L \varepsilon_a \, dx = \left[\frac{T_e + (1-2\nu)p_i A_i - (1-2\nu)p_e A_e}{EA} + \alpha t \right]_m L \qquad (5.13)$$

where all values of effective tension T_e, internal pressure p_i, external pressure p_e, and temperature t correspond to the midpoint of the pipe (denoted by external subscript "m") and α is the coefficient of axial thermal expansion. The temperature t is measured with respect to some datum.

Hence, the initial and the final values of axial stretch (e_i and e_f, respectively), before and after upending, are given by equations (5.14) and (5.15):

$$e_i = \left(\int_0^L \varepsilon_a \, dx \right)_i = \left[\frac{T_e + (1-2\nu)p_i A_i - (1-2\nu)p_e A_e}{EA} + \alpha t \right]_{mi} L \quad (5.14)$$

$$e_f = \left(\int_0^L \varepsilon_a \, dx \right)_f = \left[\frac{T_e + (1-2\nu)p_i A_i - (1-2\nu)p_e A_e}{EA} + \alpha t \right]_{mf} L \quad (5.15)$$

All the characteristics (T_e, p_i, p_e, and t) at the midpoint are known both before and after upending. Before upending, the midpoint initial effective tension $(T_e)_{mi}$ is equal to the pipe external tension, which may be zero. After upending, $(T_e)_{mf}$ is equal to the top tension minus the apparent weight of the upper half of the pipe.

The difference between the initial and final elongations gives the stretch ($\Delta e = e_f - e_i$):

$$\Delta e = \left[\frac{\Delta T_e + (1-2\nu)\Delta p_i A_i - (1-2\nu)\Delta p_e A_e}{EA} + \alpha \Delta t \right]_m L \qquad (5.16)$$

where all parameter changes (ΔT_e, Δp_i, Δp_e, and Δt) refer to the midpoint of the pipe.

For pipes made from anisotropic materials, the same approach can be used. However, the single Poisson's ratio in equations (5.13)–(5.16) has to be replaced by the equivalent Poisson's ratios $\overline{\nu_i}$ and $\overline{\nu_e}$, as in equation (5.7).

Riser Tension and Stretch Resulting from Internal Changes

Risers are frequently subject to changes of temperature, pressure, and internal fluids. All these parameter changes will influence the riser tension or axial stretch or both. It is shown below that the relationship between riser top tension T_t and stretch e can always be written in the form

$$T_t = e\, k_{riser} + \{F_w - G_{pt}\} \tag{5.17}$$

where k_{riser} is the riser axial stiffness, F_w is a function of the riser apparent weight, and G_{pt} is a function of riser pressure and temperature.

To find the influence of riser parameters (temperatures, pressures, and internal fluids) on tension and stretch, it is only necessary to calculate the changes ΔF_w and ΔG_{pt} induced by the parameter changes. Then the changes in top tension ΔT_t and stretch Δe are related by

$$\Delta T_t = \Delta e(k_{riser}) + \{\Delta F_w - \Delta G_{pt}\} \tag{5.18}$$

The functions F_w and G_{pt} and, hence, ΔF_w and ΔG_{pt} depend on the riser details. Expressions are derived in the following subsections, first for the case of a uniform riser consisting of a single tube, for which all riser characteristics are assumed to be either constant or to vary linearly over the riser length, and then for a single-tube segmented riser, for which the characteristics are assumed to be constant or linear for each riser segment. Multi-tube risers are examined subsequently. It is not possible to derive general universally applicable expressions for F_w and G_{pt} and, hence, ΔF_w and ΔG_{pt} for multi-tube risers, because of the vast number of different possible combinations. Nevertheless, the procedure to formulate expressions for ΔF_w and ΔG_{pt} for any particular multi-tube riser is explained.

Single-tube uniform risers

For a uniform near-vertical riser tube, with constant linear apparent weight and constant cross-sectional areas A_i and A_e and with characteristics of pressure and temperature varying linearly between the riser extremities, an approach similar to the pipe-upending problem can be used. The axial stretch can be calculated from the initial and final mean values of effective tension, pressure, and temperature at the midpoint of the tube.

Equation (5.13) can be rewritten as equation (5.19), where all characteristics (T_e, p_i, p_e, and t) correspond to the riser midpoint:

$$e\left(\frac{EA}{L}\right) = T_{em} + \left[(1-2\nu)p_iA_i - (1-2\nu)p_eA_e + (EA)\alpha t\right]_m \quad (5.19)$$

The midpoint effective tension T_{em} is related to the top tension T_t and the apparent weight per unit length w_a as follows:

$$T_{em} = T_t - w_a\frac{L}{2} \quad (5.20)$$

Since the tube axial stiffness k_{tube} is equal to EA/L, equation (5.19) can be rewritten as equation (5.21). (Note that in this subsection, the subscript "tube" is used instead of "riser" since the equations will later be applied to the components of multi-tube risers.)

$$T_t = e(k_{tube}) + \left\{\frac{w_aL}{2} - \left[(1-2\nu)p_iA_i - (1-2\nu)p_eA_e + (EA)\alpha t\right]_m\right\} \quad (5.21)$$

which is in the form of equation (5.17) with

$$F_w = \frac{w_aL}{2} \quad (5.22)$$

$$G_{pt} = \left[(1-2\nu)p_iA_i - (1-2\nu)p_eA_e + (EA)\alpha t\right]_m \quad (5.23)$$

Equations (5.22) and (5.23) can be formulated before and after changes to the riser internal fluid, pressure, and temperature. Taking the difference then gives

$$\Delta F_{w} = \Delta w_{a} \frac{L}{2} \qquad (5.24)$$

$$\Delta G_{pt} = \left[(1-2\nu)\Delta p_{i} A_{i} - (1-2\nu)\Delta p_{e} A_{e} + (EA)\alpha \Delta t\right]_{m} \qquad (5.25)$$

Equations (5.24) and (5.25) can be used in equation (5.18) for a uniform riser tube. Note that for a tube exposed to constant external pressure, $\Delta p_{e} = 0$.

Single-tube segmented risers

Riser characteristics of cross-section A_i and A_e and apparent weight per unit length w_a are not always constant over the entire length of the riser. Likewise, pressure and temperature gradients may change at points along a riser. Such risers can be divided into segments for which cross-section and apparent weight per unit length *are* constant and pressures and temperature *do* vary linearly between the segment extremities. For a segmented riser tube, the total stretch e has to be found by summing the elongations of all the segments ($e = \Sigma e_{seg}$).

The effective tension T_{em} at the midpoint of a particular segment is related to the top tension T_t by

$$T_{em} = T_{t} - (W_{a})_{t-m} \qquad (5.26)$$

where $(W_a)_{t-m}$ is the total apparent weight of the tube plus contents between the riser top end and the segment midpoint (hence, the subscript “t – m”). Note that for the upper segment, $(W_a)_{t-m}$ is equal to the apparent weight of half the segment. For other segments, it is equal to the apparent weight of half the segment concerned, plus the total apparent weight of all the segments above it.

In the following equations, the abbreviation $[\,]_m$ is used, as defined in equation (5.27) for pressures and temperature at the midpoint of the particular segment:

$$[\,]_{m} = \left[(1-2\nu) p_{i} A_{i} - (1-2\nu) p_{e} A_{e} + (EA)\alpha t\right]_{m} \qquad (5.27)$$

Substitution of equations (5.26) and (5.27) into equation (5.19) allows the segment stretch e_{seg} to be written in terms of the top tension T_t, as in equation (5.28).

$$e_{seg} = T_t \left(\frac{L}{EA} \right)_{seg} - \left[\frac{(W_a)_{t-m} L}{EA} \right]_{seg} + \left([\ ]_m \frac{L}{EA} \right)_{seg} \tag{5.28}$$

Inside the three sets of parentheses in equation (5.28), all values including length L refer to the segment. Equation (5.28) can be written for all the segments, and the results can be summed, to give the tube total stretch e_{tube}:

$$e_{tube} = T_t \sum \left(\frac{L}{EA} \right) - \sum \left[\frac{(W_a)_{t-m} L}{EA} \right] + \sum \left([\ ]_m \frac{L}{EA} \right) \tag{5.29}$$

The axial stiffness (k_{tube}) of a tube composed of segments is equal to the reciprocal of the sum of the flexibilities (L/EA) of the segments:

$$k_{tube} = \left[\sum \left(\frac{L}{EA} \right)_{seg} \right]^{-1} \tag{5.30}$$

Multiplying through equation (5.29) by equation (5.30) and rearranging gives

$$T_t = e_{tube} k_{tube} + \left\{ k_{tube} \sum \left[\frac{(W_a)_{t-m} L}{EA} \right] - k_{tube} \sum \left([\ ]_m \frac{L}{EA} \right) \right\} \tag{5.31}$$

Equation (5.31) is in the form of equation (5.17) with

$$F_w = k_{tube} \sum \left[\frac{(W_a)_{t-m} L}{EA} \right] \tag{5.32}$$

$$G_{pt} = k_{tube} \sum \left\{ \left[(1-2\nu) p_i A_i - (1-2\nu) p_e A_e + (EA) \alpha t \right]_m \frac{L}{EA} \right\} \tag{5.33}$$

where Σ implies the sum for all segments. Equations (5.32) and (5.33) can be formulated before and after the parameter changes. The differences then give

$$\Delta F_{\mathrm{w}} = k_{\mathrm{tube}} \sum \left[\frac{(\Delta W_{\mathrm{a}})_{\mathrm{t-m}} L}{EA} \right] \tag{5.34}$$

$$\Delta G_{\mathrm{pt}} = k_{\mathrm{tube}} \sum \left\{ \left[(1-2\nu)\Delta p_{\mathrm{i}} A_{\mathrm{i}} - (1-2\nu)\Delta p_{\mathrm{e}} A_{\mathrm{e}} + (EA)\alpha \Delta t \right]_{\mathrm{m}} \frac{L}{EA} \right\} \tag{5.35}$$

Equations (5.34) and (5.35) can be used in equation (5.18) for a segmented riser tube. Note again that for a tube exposed to constant external pressure, $\Delta p_{\mathrm{e}}=0$.

Multi-tube risers

Risers are frequently composed of several barriers and several tubings, which may or may not be fitted with expansion joints. The riser outer casing may be fitted with buoyancy modules over part of its length. The cross-sectional areas A_{i} and A_{e} may include step changes. Thermal gradients may not be constant over the entire length.

Plainly, there are a large number of possible combinations that can be imagined for a multi-tube riser, and it is not possible to derive general expressions that will cover them all. However, it is possible to define procedures for obtaining F_{w}, G_{pt}, ΔF_{w}, and ΔG_{pt} for any particular case. This is the purpose of the following subsection, which describes a particular application. Extrapolation to more complicated cases should then be straightforward.

The procedure consists of decomposing the effective tension (and, hence, top tension) and the apparent weight of the riser into components corresponding to each tube. Equations relating the top tension T_{t} to the riser stretch *e,* similar to equation (5.17), can then be formulated for each tube and summed to give the version of equation (5.17) that applies to the complete riser. This is done for the riser in its initial condition and is then repeated for the final condition, after the parameter changes.

Taking the difference again leads to a formula resembling equation (5.18), which allows the influence of the parameter changes to be deduced. When the apparent weight is decomposed into its components for each tube, the revised definition of apparent weight given in chapter 3 for tubes within tubes must be carefully respected (see the bullet points emphasized following equation [3.13]).

The relationship between the tube top-tension component $(T_t)_{tube}$ and the tube stretch e has already been derived for uniform tubes (see equation [5.21]) and for segmented tubes (see equation [5.31]). A tube with an *expansion joint* will also have a component of top tension, which must be included when formulating the total riser top tension, although that top-tension component will be independent of the tube stretch.

Top tension of tubing with expansion joint. Tubing with an expansion joint has zero axial stiffness, but it has a top-tension component. Since the effective tension in such a tube is known at the expansion joint, the tube top-tension component can be calculated.

The effective tension at an expansion joint with a simple O-ring seal is given by

$$T_e = -p_i A_j + p_e A_j \tag{5.36}$$

where A_j is the seal enclosed area (cf. equation [3.5] for a lugged connector, but without lugs).

For tubing of length d (between the top end and the expansion joint) and apparent weight per unit length w_a, the top-tension component will be given by

$$(T_t)_{tube} = w_a d - \left[p_i A_j - p_e A_j\right]_{EJ} \tag{5.37}$$

where subscript "EJ" refers to details of the expansion joint (pressures and seal area).

Application of multi-tube vertical risers. The procedures to formulate F_w and G_{pt} for a multi-tube riser are best explained by applying them to a particular riser. Extrapolation to other cases should then be straightforward.

For simplicity, just one case will be considered closely consisting of a double-barrier riser with dual completion. The outer riser barrier (tube 1) comprises a change of section and is hence segmented, whereas the inner riser barrier (tube 2) is uniform. The tubings are both uniform, but one (tube 3) is fixed at the bottom end of the well, while the other (tube 4) is equipped with an expansion joint in the well.

Figure 5–4 shows a sketch of the riser and its components. The lengths of the riser casings and the tubings are different, but they all experience the same stretch *e*.

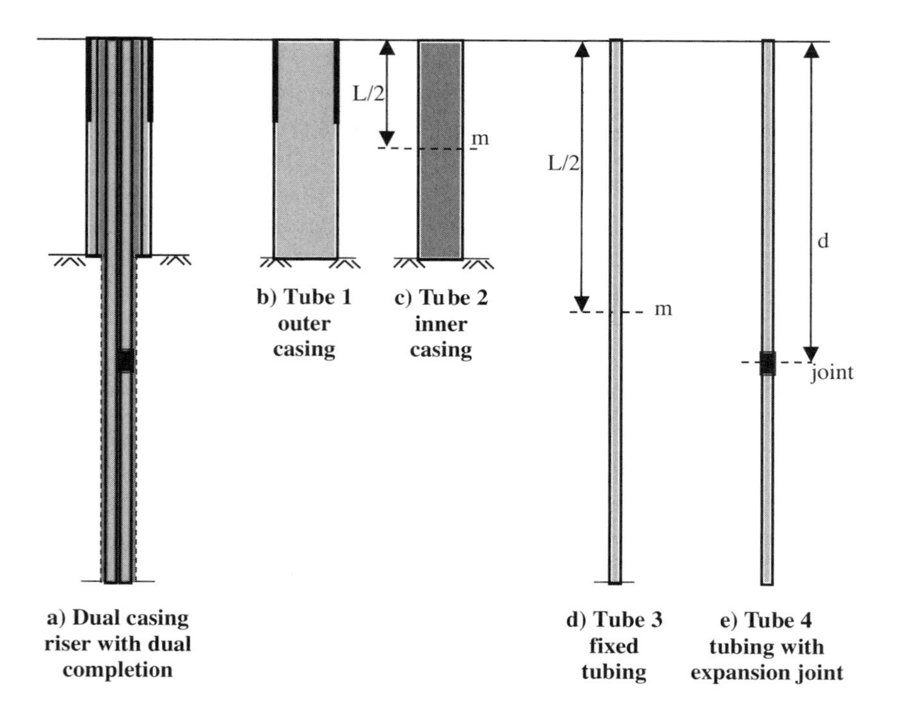

Fig. 5–4. *Dual casing riser with dual completion, showing schematic composition*

The top-tension contribution from each tube and its relation to the riser axial stretch *e* can be obtained from the equations already derived for single-tube risers: equation (5.31), for the segmented casing (tube 1); equation (5.21), for the uniform tubes (2 and 3); and equation (5.37), for the tube with an expansion joint (tube 4). The equations can be summed to give the relationship between the total top tension T_t and the axial stretch *e*.

By use of subscripts 1–4 for the tubes shown in figure 5–4, the equations relating the top-tension component of each tube to the axial stretch are given by equations (5.38)–(5.41). (The abbreviation $[\ ]_m$ used in equation [5.38] was defined in equation [5.27].)

$$T_{t1} = e(k_1) + k_1 \sum \left[\frac{(W_a)_{t-m} L}{EA} \right]_1 - k_1 \sum \left([\,]_m \frac{L}{EA} \right)_1 \qquad (5.38)$$

$$T_{t2} = e(k_2) + \left(w_a \frac{L}{2} \right)_2 - [(1-2\nu) p_i A_i - (1-2\nu) p_e A_e + (EA)\alpha t]_{m2} \qquad (5.39)$$

$$T_{t3} = e(k_3) + \left(w_a \frac{L}{2} \right)_3 - [(1-2\nu) p_i A_i - (1-2\nu) p_e A_e + (EA)\alpha t]_{m3} \qquad (5.40)$$

$$T_{t4} = \quad (w_a d)_4 - [p_i A_J - p_e A_J]_{EJ4} \qquad (5.41)$$

The sum of all the top-tension components is the riser top tension T_t. Equation (5.42) is the sum of equations (5.38)–(5.41):

$$T_t = e(k_{riser}) + \{F_w - G_{pt}\} \qquad (5.42)$$

where the stiffness k_{riser} is the sum of the stiffnesses of all the fixed tubes (Σk_{tubes}), F_w is the sum of all the terms relating to apparent weight of the tubes, and G_{pt} is the sum of all the terms relating to pressure and temperature.

Equation (5.42) is therefore in the form of equation (5.17). It can be evaluated before and after parameter changes. Taking the difference then gives the required values of ΔF_w and ΔG_{pt} in equation (5.18).

Influence of Tensioners

It has been shown that the relationship between a change of top tension ΔT_t and a change of riser stretch Δe resulting from riser internal parameter changes can always be written in the form of equation (5.18). If the riser is equipped with tensioners of stiffness k_{tens} and the overall length of the riser plus tensioners remains unchanged, then the tensioners must contract by an amount equal to the riser stretch. Hence, the following relationship between ΔT_t and Δe must also be respected:

$$\Delta T_t = -k_{tens} \Delta e \qquad (5.43)$$

Calling k_{r+t} the stiffness of the combined system, riser plus tensioners,

$$k_{r+t} = \left(\frac{1}{k_{riser}} + \frac{1}{k_{tens}} \right)^{-1} = \frac{k_{riser} k_{tens}}{k_{riser} + k_{tens}} \tag{5.44}$$

Elimination of Δe between equations (5.18) and (5.43) leads to the first equality of equation (5.45). Substitution from equation (5.44) then leads to the second equality:

$$\Delta T_t = \frac{\{\Delta F_w - \Delta G_{pt}\}}{k_{riser} + k_{tens}} k_{tens} = \{\Delta F_w - \Delta G_{pt}\} \frac{k_{r+t}}{k_{riser}} \tag{5.45}$$

From equation (5.43) and the first equality of equation (5.45),

$$\Delta e = -\frac{\Delta T_t}{k_{tens}} = -\frac{\{\Delta F_w - \Delta G_{pt}\}}{k_{riser} + k_{tens}} \tag{5.46}$$

If the riser is locked off ($k_{tens} = \infty$), $k_{r+t} = k_{riser}$. Then, $\Delta e = 0$, and from equation (5.45),

$$\Delta T_t = \{\Delta F_w - \Delta G_{pt}\} \tag{5.47}$$

If the riser is equipped with perfect tensioners with zero stiffness ($k_{tens} = 0$), top tension is constant ($\Delta T_t = 0$), and the riser stretch is given by

$$\Delta e = -\frac{\{\Delta F_w - \Delta G_{pt}\}}{k_{riser}} \tag{5.48}$$

Note from the second equality of equation (5.45), that the change in top tension ΔT_t is equal to the tension change of the locked-off riser $\{\Delta F_w - \Delta G_{pt}\}$, times the ratio k_{r+t}/k_{riser}.

Summary

Strain equations have been derived in terms of the true wall tension T_{tw} and in terms of the effective tension T_e, for both isotropic and anisotropic pipes. Equivalent Poisson's ratios $\overline{\nu}_i$ and $\overline{\nu}_e$, required in the calculation of pressure-induced axial strains of anisotropic pipes, have been defined. Equations relating $\overline{\nu_i}$ and $\overline{\nu_e}$ to the material characteristics have been presented, and simple ways in which they can be determined by tests on tube samples have been explained.

Several particular problems have been discussed in detail. These include pressure-induced buckling of tubes for which axial expansion is prevented; pipe or riser stretch resulting from upending; stretch of single-tube risers (uniform or segmented), resulting from changes of temperature, pressure, and internal fluid; and stretch of multi-tube risers, resulting from changes of temperatures, pressures, and internal fluid densities. The influence of riser-tensioner stiffness on the riser stretch and tension change resulting from internal parameter changes has also been discussed.

All the important relationships between pipe/riser behavior and effective tension have been established in chapters 2–5. The static behavior of complete risers can now be studied. However, before that, much can be learned from the analytical study of the behavior of simple tensioned beams. This is the subject of the next chapter.

References

[1]Sparks, C. P. 1984. The influence of tension, pressure and weight on pipe and riser deformations and stresses. *Transactions of ASME. Journal of Energy Resources Technology*. 106 (March), 46–54.

[2]Sparks, C. P., P. Odru, H. Bono, and G. Metivaud. 1988. Mechanical testing of high-performance composite tubes for TLP production risers. Paper OTC 5797, presented at the Offshore Technology Conference, Houston;
Williams, J. G. 1991. Developments in composite structures for the offshore industry. Paper OTC 6579, presented at the Offshore Technology Conference, Houston;
Johnson, D. B. 1999. Rigid composite risers for deepwater oil production. Paper presented at the Intertech 2nd International Conference on the Global Outlook for Carbon Fibers, San Diego;
Salama, M. M., G. Stjern, T. Storhaug, B. Spencer, and A. Echtermeyer. 2002. The first offshore field installation for a composite riser joint. Paper OTC 14018, presented at the Offshore Technology Conference, Houston.

[3]Sparks, C. P., and J. Schmitt. 1990. Optimized composite tubes for riser applications. Paper presented at the Offshore Mechanics and Arctic Engineering Conference, Houston.

[4]Odru, P., and C. P. Sparks. 1991. Thick walled composite tubes: Calculation and measured behavior. Paper presented at the 9th Offshore Mechanics and Arctic Engineering Conference, Stavanger, Norway.

[5]Jones, R. M. 1975. *Mechanics of Composite Materials.* McGraw-Hill, New York;
Tsai, S. W. 1986. *Composite Designs*. Think Composite, Dayton, Ohio

[6]Ibid.

[7]Sparks and Schmitt, 1990.

[8]Palmer, A. C., and J. A. S. Baldry. 1974. Lateral buckling of axially constrained pipelines. *Journal of Petroleum Technology.* November, 1283–1284.

[9]Poirette, Y., G. Guesnon, and D. Dupuis. 2006. First hyperstatic riser joint field test for deep offshore drilling. Paper IADC/SPE SPE-99005-PP, presented at the IADC/SPE Drilling Conference, Miami.

[10]Melve, B. 2005. Subsea spoolable composite pipes: lessons learnt during the Asgard Project. Paper presented at the 4th International Conference on Composite Materials and Structures for Offshore Operations, Houston.

6

Tensioned-Beam Behavior

This chapter uses analytical methods to study the influence of bending stiffness on tensioned-beam behavior. Results obtained here are applied in subsequent chapters to near-vertical risers, stress joints, riser bundles, catenary risers, and riser transverse vibrations.

Tensioned-beam deflections and curvature are governed by the fourth-order differential equation derived in the first section of appendix A (see equation [A.6]). For the case with constant bending stiffness, the equation can be written as follows:

$$EI\frac{d^4y}{dx^4} - T\frac{d^2y}{dx^2} - w\frac{dy}{dx} - f(x) = 0 \qquad (6.1)$$

where EI is the bending stiffness, T is the effective tension, w is the apparent weight per unit length, and $f(x)$ is the applied load function. For the special case with constant tension (i.e., for $w = 0$), equation (6.1) becomes

$$EI\frac{d^4y}{dx^4} - T\frac{d^2y}{dx^2} - f(x) = 0 \qquad (6.2)$$

Equation (6.2) is the standard equation of a horizontal tensioned beam subject to load function $f(x)$, where x is the horizontal axis and y the vertical axis, as shown in figure 6–1. Much can be learned about riser behavior by studying the solution to equation (6.2), since it can be solved analytically.[1]

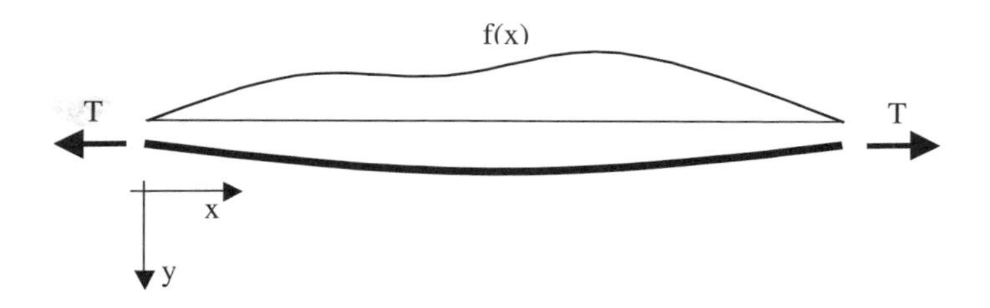

Fig. 6–1. *Horizontal tensioned beam*

In this chapter, for simplicity and clarity, the word *beam* is used for cases in which the bending stiffness *EI* is included in the analysis. The word *cable* is used for cases in which *EI* is neglected.

Excel File *Tensioned-Beam.xls*

The Excel file *Tensioned-Beam.xls* can be used to explore and verify relationships and trends mentioned in this chapter using different data. It contains three work sheets. The *Curvature* work sheet allows the curvature of beams with and without bending stiffness to be compared. The *Stiffness* work sheet compares the angular stiffness, when the beam is subject to an end moment, with that given by simple approximate analytical formulae. The *Angles* work sheet compares the beam end angles with those deduced from cable end angles calculated by ignoring bending stiffness.

Influence of Bending Stiffness for Beams with Uniform Load

The influence of bending stiffness on a tensioned beam can be understood most clearly by first considering the beam subject to uniform load (f_0). The resultant equations of beam deflection and beam curvature can be found by solving equation (6.2). For the case with zero end moment, this yields the following expression for the curvature $1/R$:

$$\left(\frac{1}{R}\right)_x = -\frac{d^2y}{dx^2} = \frac{f_0}{T}\left(1 - \frac{\sinh k(L-x) + \sinh kx}{\sinh kL}\right) \tag{6.3}$$

where the parameter k is given by

$$k = \sqrt{\frac{T}{EI}} \tag{6.4}$$

Parameter k has the dimensions of length^{-1}. Hence, kL is a dimensionless *flexibility factor* that increases as EI decreases. Equation (6.3) is traced on figure 6–2 for three values: kL = 10, 20, and 60.

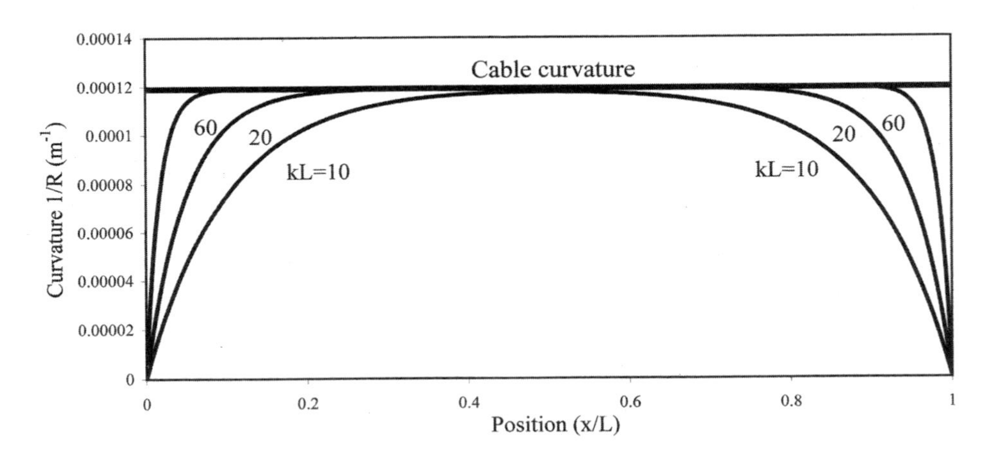

Fig. 6–2. *Curvature of a tensioned beam under uniform load*

As seen in figure 6–2, the *beam curvature* is asymptotic to the *cable curvature* over a central length, the extent of which depends on the value of the flexibility factor kL. This can be explored using the *Curvature* work sheet of *Tensioned-Beam.xls* for different values of uniform load ($f_0 = f_L$), length, tension, and kL (with end moments put equal to zero). The work sheet compares beam and cable curvature both graphically and in tabular form.

Since the object of the work sheet is to show the influence of the flexibility factor kL, this is input as data. The corresponding value of EI is then calculated by the file.

The tendency of the beam curvature to converge to the cable curvature, as kL increases, can be understood from equation (6.3), which can be rewritten as

$$\left(\frac{1}{R}\right)_x = -\left[\frac{f_0}{T}\right]\frac{\text{Sh}\,k(L-x)}{\text{Sh}\,kL} + \left[\frac{f_0}{T}\right] - \left[\frac{f_0}{T}\right]\frac{\text{Sh}\,kx}{\text{Sh}\,kL} \qquad (6.5)$$

where the abbreviation "Sh" is used to denote sinh.

As kL increases, the ratio Sh $k(L - x)$/(Sh kL) converges to e^{-kx}, and Sh kx/(Sh kL) converges to $e^{-k(L-x)}$. Equation (6.5) then becomes

$$\left(\frac{1}{R}\right)_x = -\left[\frac{f_0}{T}\right]e^{-kx} + \left[\frac{f_0}{T}\right] - \left[\frac{f_0}{T}\right]e^{-k(L-x)} \qquad (6.6)$$

Thus, beam curvature is composed of three terms. The effects of the left and right supports (at $x = 0$ and $x = L$) are expressed by the first and last terms, respectively, on the right-hand side of equations (6.5) and (6.6), and these effects dissipate exponentially with increasing distance from the supports.

For $kL > 10$, there is then a central zone, where the influence of the supports is negligible and the beam curvature $1/R$ is equal to the cable curvature (f_0/T) given by the middle bracketed term on the right-hand side of equation (6.6). As kL increases, the length of this central zone increases, and the zones close to the supports, where the beam ends straighten locally, shorten.

Influence of Bending Stiffness for Beams with Parabolic Load

For beams under *constant tension,* the tendency for beam curvature to converge very closely to cable curvature, except for zones close to the supports, is also true for more complicated loadings. Analytical expressions similar to equations (6.5) and (6.6) can always be formulated, whatever the loading, although the complexity of the three bracketed terms will depend on the complexity of the load function (see appendix F).

A *parabolic load function,* with loads f_0 and f_L at the extremities and with $\sqrt{\text{load}}$ varying linearly between them, is interesting since it is similar to the load induced by a shear current with constant drag coefficient C_D. Such a load function is given by equation (6.7):

$$f_x = \left[\sqrt{f_0} + \left(\sqrt{f_L} - \sqrt{f_0}\right)\frac{x}{L}\right]^2 \qquad (6.7)$$

If the beam is also subject to end moments, M_0 (at $x = 0$) and M_L (at $x = L$), then functions G_0 and G_L can be defined as follows:

$$G_0 = \frac{M_0}{EI} - \frac{f_0}{T} - \frac{2}{T}\left(\frac{\sqrt{f_L} - \sqrt{f_0}}{kL}\right)^2 \qquad (6.8)$$

$$G_L = \frac{M_L}{EI} - \frac{f_L}{T} - \frac{2}{T}\left(\frac{\sqrt{f_L} - \sqrt{f_0}}{kL}\right)^2 \qquad (6.9)$$

The analytical solution to equation (6.2) then gives the beam curvature as follows (see equation (F.15) of appendix F):

$$\left(\frac{1}{R}\right)_x = G_0 \frac{\operatorname{Sh} k(L-x)}{\operatorname{Sh} kL} + \left[\frac{f_x}{T} + \frac{2}{T}\left(\frac{\sqrt{f_L} - \sqrt{f_0}}{kL}\right)^2\right] + G_L \frac{\operatorname{Sh} kx}{\operatorname{Sh} kL} \qquad (6.10)$$

As kL increases, equation (6.10) tends toward

$$\left(\frac{1}{R}\right)_x = G_0 e^{-kx} + \left[\frac{f_x}{T} + \frac{2}{T}\left(\frac{\sqrt{f_L} - \sqrt{f_0}}{kL}\right)^2\right] + G_L e^{-k(L-x)} \qquad (6.11)$$

The effects of the end supports and the end moments, respectively expressed by the first and last terms on the right-hand side of equations (6.10) and (6.11), again dissipate exponentially with distance from the supports.

Since $e^{-kx} = 0.05$ and 0.01 for $kx = 3$ and 4.6, respectively, it follows that the effects of the supports are reduced to 5% and 1% of their end values at distances $x/L = 3/kL$ and $4.6/kL$, respectively, from the supports. This can be appreciated from the example of figure 6–3 for a particular case with parabolic loading and large end moments, for which $kL = 20$.

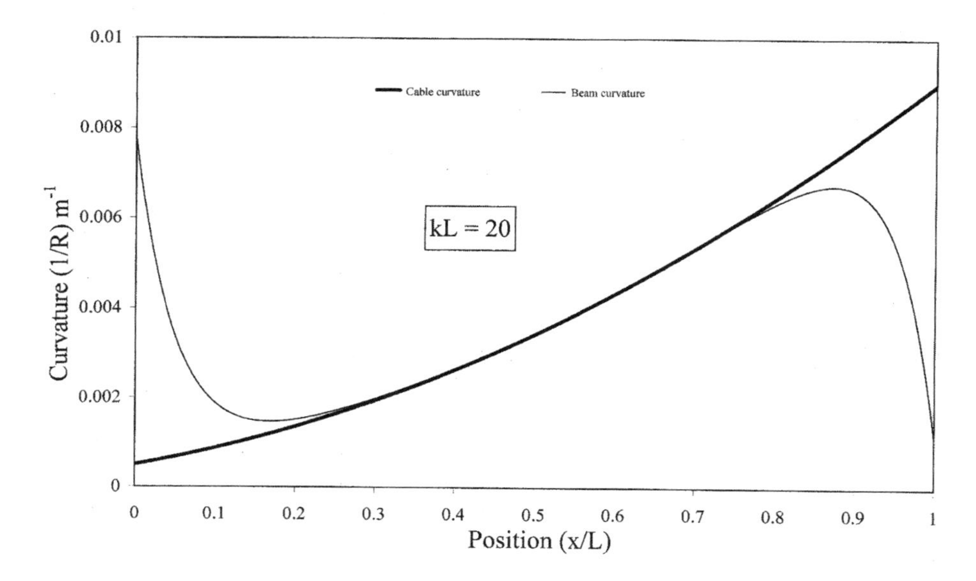

Fig. 6–3. *Tensioned beam under parabolic load with end moments*

The reader can explore the aforementioned effects more fully, with different data, using the *Curvature* work sheet of *Tensioned-Beam.xls.* In this work sheet, the beam curvature, given by equation (6.10), is compared with the cable curvature (f_x/T).

For the central zone, where the effects of the extremities (with or without moments) become negligible, the curvature is given by the middle term on the right-hand side of equation (6.11):

$$\left(\frac{1}{R}\right)_x = \frac{f_x}{T} + \frac{2}{T}\left(\frac{\sqrt{f_L} - \sqrt{f_0}}{kL}\right)^2 \qquad (6.12)$$

The curvature $1/R$ in the central zone, as given by equation (6.12), is close to—but slightly greater than—the cable curvature f_x/T. The difference, given by equation (6.13), is also output by the *Curvature* work sheet and becomes increasingly negligible as kL increases:

$$\left(\frac{1}{R}\right)_{\text{beam}} - \left(\frac{1}{R}\right)_{\text{cable}} = \frac{2}{T}\left(\frac{\sqrt{f_L} - \sqrt{f_0}}{kL}\right)^2 \qquad (6.13)$$

The small increase in curvature for the beam, as compared to the cable, is the result of a slight redistribution of the effect of *nonuniform load,* owing to the shear capacity of the beam. It can be easily understood by considering the curvature of tensioned cables and beams subject to a load over only *part* of their length. The tensioned cable will be curved only below the load. To either side of the load, the cable will be straight. By contrast, a tensioned beam, as a result of its shear capacity, will be curved over its entire length.

Note, if beam end moments (M_0 and M_L) are deliberately chosen to give $G_0 = G_L = 0$, the effects of the end supports are eliminated. The difference in curvature given by equation (6.13) will then be *constant* along the entire length of the beam. The required moments (M_0 and M_L) to give $G_0 = G_L = 0$ are provided by the work sheet. They are close to $M_0 = EI f_0 / T$ and $M_L = EI f_L / T$ (see eqq. [6.8] and [6.9]), which are also given by the work sheet.

Influence of End Moment

If an end moment M_0 is applied to a tensioned beam, as in figure 6–4, the resultant equations of beam deflection, slope, and curvature can be found by solving equation (6.2).

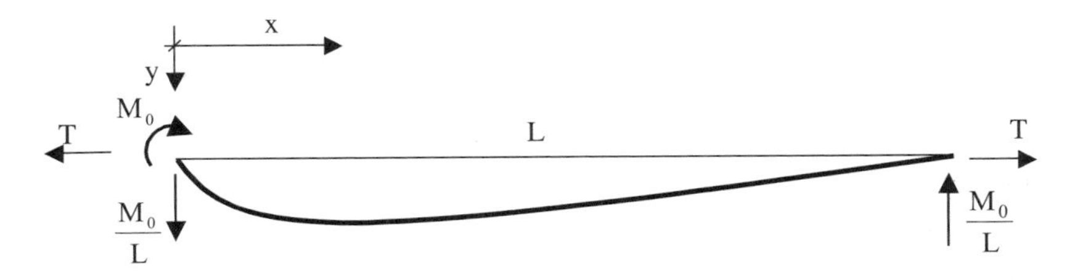

Fig. 6–4. *Tensioned beam with applied end moment*

This yields the following expressions for the beam slope dy/dx and curvature $1/R$ as a function of x:

$$\left(\frac{dy}{dx}\right)_x = \frac{kM_0}{T}\left[\frac{\cosh k(L-x)}{\sinh kL} - \frac{1}{kL}\right] \tag{6.14}$$

$$\left(\frac{1}{R}\right)_x = -\frac{d^2y}{dx^2} = \frac{M_x}{EI} = \frac{M_0}{EI}\frac{\sinh k(L-x)}{\sinh kL} \tag{6.15}$$

The curvature given by equation (6.15) agrees with that given by equation (6.10) for zero loads and zero moment M_L (i.e., for $f_0 = f_L = M_L = 0$). The decreasing effect of end moment with distance from the extremity to which it is applied is shown on figure 6–5.

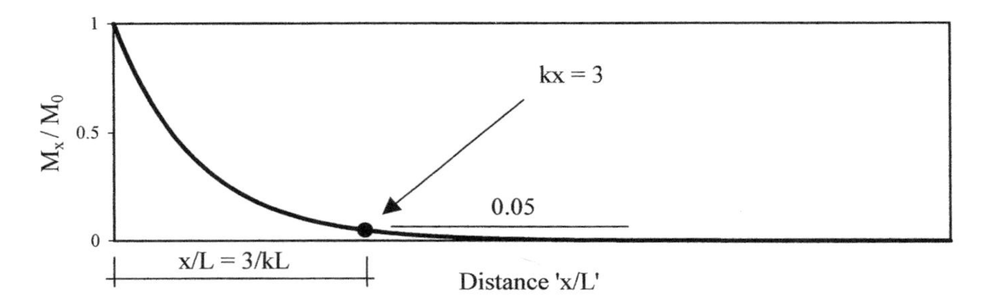

Fig. 6–5. *Effect of end moment M_0 on a tensioned beam*

As kL increases, equation (6.15) tends toward equation (6.16), which agrees with equation (6.11) for $f_0 = f_L = M_L = 0$:

$$\left(\frac{1}{R}\right)_x = \frac{M_x}{EI} = \frac{M_0}{EI} e^{-kx} \tag{6.16}$$

The moment M_x decreases exponentially with distance from the beam end to which it is applied, in a similar fashion to the decrease of the curvature in the previous section. The moment M_x is reduced to 5% of the end moment M_0 for $kx = 3$ and to 1% for $kx = 4.6$.

Note also from equation (6.14) that as kL increases, the angle at the far extremity ($x = L$) tends toward a value independent of the stiffness: $-M_0/TL$.

End Rotational Stiffness

From equation (6.4), $T / k = \sqrt{TEI}$; thus, equation (6.14) can be used to give the *exact value* of the end rotational stiffness (moment/radian):

$$M_0 \bigg/ \left(\frac{dy}{dx}\right)_0 = \sqrt{TEI}\left(\frac{1}{\tanh kL} - \frac{1}{kL}\right)^{-1} \tag{6.17}$$

For $kL > 3$, $\tanh kL = 1$ to within 0.5%. Equation (6.17) is then closely approximated by

$$M_0 \bigg/ \left(\frac{dy}{dx}\right)_0 = \sqrt{TEI}\left(1 - \frac{1}{kL}\right)^{-1} \qquad (6.18)$$

For larger values of kL, equation (6.18) can be further approximated to

$$M_0 \bigg/ \left(\frac{dy}{dx}\right)_0 = \sqrt{TEI} \qquad (6.19)$$

The accuracy of the preceding approximations for the end rotational stiffness can be verified using the *Stiffness* work sheet of *Tensioned-Beam.xls*.

Table 6–1. *Example of end rotational stiffness and end shear (for kL = 20)*

End rotational stiffness	kN-m/radian	Precision ratio
Exact value (M / θ)	31,579	-
$\sqrt{[T.EI]}$ / [1 - 1/kL]	31,579	1.00
$\sqrt{[T.EI]}$	30,000	0.95
End shear	**kN**	**Precision ratio**
Exact shear force	-66.67	-
$-k.M_0$	-66.67	1.00

The upper part of table 6–1 shows an example of the end rotational stiffness as output by the *Stiffness* work sheet for a case for which kL = 20. The work sheet outputs the exact stiffness, as given by equation (6.17), and compares it with the approximate values given by equations (6.18) and (6.19). (The lower part of table 6–1 is discussed in the next section.)

End Shear Force

The shear force F_x associated with an end moment can be found by differentiating equation (6.15), which gives

$$F_x = \frac{dM}{dx} = \frac{d}{dx}\left(-EI\frac{d^2y}{dx^2}\right) = -kM_0\frac{\cosh k(L-x)}{\sinh kL} \qquad (6.20)$$

Therefore, the end shear force F_0 (for $x = 0$) is given *exactly* by

$$F_0 = -kM_0 \frac{1}{\tanh kL} \tag{6.21}$$

For $kL > 3$, equation (6.21) can be approximated to

$$F_0 = -kM_0 \tag{6.22}$$

The accuracy of equation (6.22) can also be verified using the *Stiffness* work sheet of *Tensioned-beam.xls*. The lower part of table 6–1 shows an example of the end shear force as given by the work sheet for a case with $kL = 20$, where the exact value of the end shear force given by equation (6.21) is compared with $-kM_0$.

Note that the shear force F_0 acts perpendicularly to the beam axis. It is related to the end reaction V_0, which acts perpendicularly to the x-axis, by

$$V_0 = F_0 + T\frac{dy}{dx} = -\frac{M_0}{L} \tag{6.23}$$

Beam Angles Deduced from Cable Angles

Since beam curvature is close to cable curvature except for zones close to the supports, it should be possible to evaluate the *beam end angles* with reasonable precision from the *cable end angles.* This could be achieved by applying a correction to take into account the local straightening at the beam extremities, resulting from the influence of bending stiffness.

Beam end moments can be applied to force the beam to experience the same end curvature as the cable (i.e., f_0/T at $x = 0$ and f_L/T at $x = L$). The required beam end moments are given by equations (6.24) and (6.25):

$$M_0 = EI\frac{f_0}{T} \tag{6.24}$$

$$M_L = EI\frac{f_L}{T} \tag{6.25}$$

When the end moments obtained in equations (6.24) and (6.25) are applied, the whole beam profile should closely match the cable profile. Since the end rotational stiffness (M_0 / θ_0) is known, the changes in the end angles that occur when the end moments are released can be estimated.

The end rotational stiffness was given approximately by $\sqrt{TEI}$ (see equation [6.19]). Hence the changes in the end angles, as the end moments are released, can be obtained approximately by dividing equations (6.24) and (6.25) by that stiffness giving:

$$\delta\theta_0 = \frac{EI\, f_0}{T\sqrt{TEI}} = \frac{f_0}{kT} \tag{6.26}$$

$$\delta\theta_L = \frac{EI\, f_L}{T\sqrt{TEI}} = \frac{f_L}{kT} \tag{6.27}$$

The differences between the cable and the beam end angles are then

$$(\theta_{\text{cable}} - \theta_{\text{beam}})_0 = \frac{f_0}{kT} \tag{6.28}$$

and

$$(\theta_{\text{cable}} - \theta_{\text{beam}})_L = -\frac{f_L}{kT} \tag{6.29}$$

The accuracy of these approximations improves as kL increases, as the reader can verify with different data, using the *Angles* work sheet of *Tensioned-beam.xls*.

Table 6–2. *Example of beam angles calculated from cable angles (for kL = 20)*

Position	x=0	x=L
Cable angles	9.55°	-5.98°
Beam angles	8.24°	-5.64°
Beam angles obtained from cable angles by correcting for EI		
Beam angles	8.17°	-5.67°
Precision ratio	0.99	1.01

The upper part of table 6–2 shows the cable angles and the beam angles calculated analytically by the program, for an example with $kL = 20$. The lower part of table 6–2 gives the beam angles deduced from the cable angles, using equations (6.28) and (6.29).

Summary

Analytical methods have been used to explore the behavior of beams under *constant tension* and to compare this with the behavior of tensioned cables under the same load. This has led to a number of simple observations and expressions.

The behavior of beams and cables has been compared for cases subject to uniform loads and to parabolic loads. It has been shown that beam curvature is very close to cable curvature (for which EI is neglected) everywhere except for zones close to the supports. The extent of those zones depends on the flexibility factor kL, where k is defined in equation (6.4).

The effect of end moments has also been seen to dissipate exponentially with distance from the end supports, as a function of the flexibility factor kL. Simple expressions have been derived for the end rotational stiffness (equations [6.18] and [6.19]) and for the shear force associated with an end moment (equation [6.22]). It has further been shown that beam end angles can be deduced with reasonable accuracy from cable end angles by allowing for the local straightening at the beam ends.

The work sheets of the Excel file *Tensioned-Beam.xls* have been presented. They allow the reader to verify the precision of the derived expressions with different data.

The eventual aim is to determine whether the expressions derived in this chapter can be applied or adapted to *near-vertical risers*, with varying tension. Before exploring that subject, the behavior of *near-vertical cables* must first be examined. This is the subject of the next chapter.

Reference

[1]Sparks, C. P. 1980. Mechanical behavior of marine risers—mode of influence of principal parameters. *Journal of Engineering Resources Technology, ASME*. 102 (December), 214–222.

7

Statics of Near-Vertical Cables

This chapter explores the solution to the riser equation for the case in which tension varies along the length but the bending stiffness is neglected ($EI = 0$). For simplicity, the corresponding results are referred to as *cable* results.

Once the bending stiffness has been neglected, the differential equation governing the static riser profile (see equation [6.1]) becomes

$$T\frac{d^2y}{dx^2}+w\frac{dy}{dx}+f(x)=0 \tag{7.1}$$

where T is the effective tension and w is the apparent weight. Equation (7.1) can be rewritten as equation (7.2), which can be solved analytically without difficulty:

$$\frac{d}{dx}\left(T\frac{dy}{dx}\right)+f(x)=0 \tag{7.2}$$

Uniform Cable with Current Load

Figure 7–1*a* shows a near-vertical cable with constant apparent weight per unit length, a top end lateral offset y_t, and a lateral current load function $f(x)$.

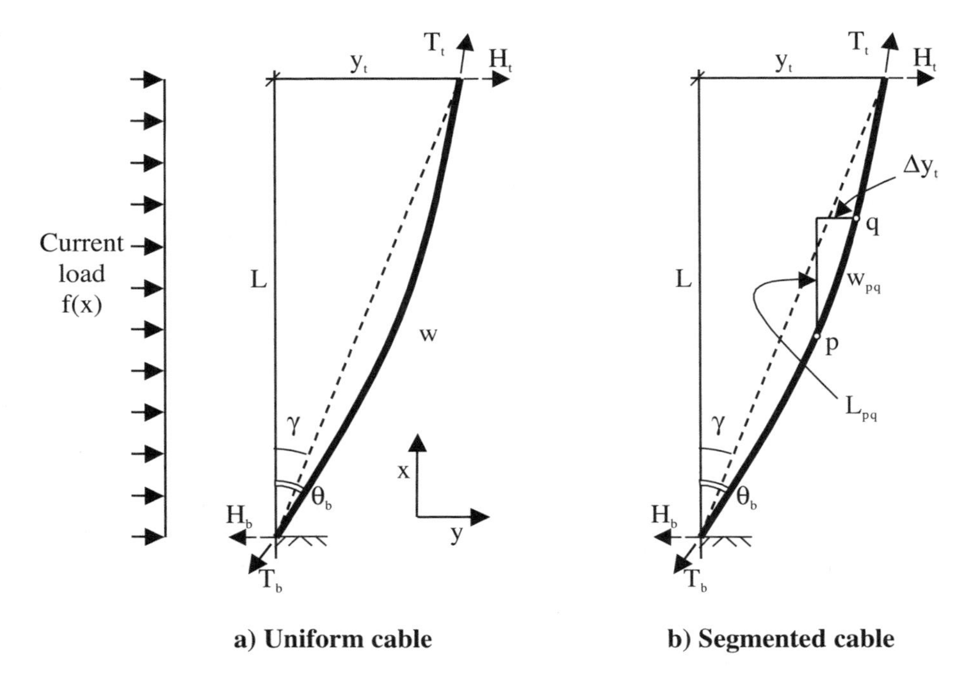

Fig. 7–1. *Near-vertical cables with current load*

For such a cable, equation (7.2) can be integrated to give

$$\frac{dy}{dx} = \frac{H}{T} = \frac{H_b - \int_0^x f(x)\,dx}{T} \tag{7.3}$$

where H is the horizontal component of the effective tension T at height x and H_b is the horizontal force at the cable bottom end.

Equation (7.3) can be used to find the slope of the cable at all points once H_b has been found. Further integration of equation (7.3) gives

$$y = \frac{H_b}{w} \ln\left(\frac{T}{T_b}\right) - \int_0^x \frac{\int_0^x f(x)\,dx}{T}\,dx \tag{7.4}$$

where y is the lateral offset with respect to the cable bottom end. Since the top-end offset y_t is known, equation (7.4) can be used to find the horizontal force H_b at the bottom end:

$$H_{\mathrm{b}} = \frac{y_{\mathrm{t}} + \int_0^L \frac{\int_0^x f(x)\,dx}{T}\,dx}{\frac{1}{w}\ln\left(\frac{T_{\mathrm{t}}}{T_{\mathrm{b}}}\right)} \tag{7.5}$$

Equations (7.3)–(7.5) allow the cable profile to be calculated at all points and the angles at the extremities to be found. The bottom-end angle, for example, is given by

$$\theta_{\mathrm{b}} = \frac{H_{\mathrm{b}}}{T_{\mathrm{b}}} \tag{7.6}$$

The preceding equations show that the law of superposition applies to the near-vertical cable. From equation (7.5), the bottom-end reaction H_{b} has two components: a component due to top-end offset y_{t} and a component due to the current load. The two combine linearly. Likewise, the components of end angles, the components of lateral offset, and the components of curvature can be calculated separately and linearly combined.

The top-end setdown (vertical movement) can also be calculated analytically from equation (7.7), although its analytical evaluation may be difficult for complicated current load functions:

$$\text{setdown} = \frac{1}{2}\int_0^L \left(\frac{dy}{dx}\right)^2 dx \tag{7.7}$$

Uniform Cable with Zero Current Load

If there is no current load, the preceding equations become very simple. The horizontal component of tension is constant along the cable ($H_{\mathrm{b}} = H_{\mathrm{t}} = H$) and is given by equation (7.8). Top and bottom angles, the top-end setdown, and the lateral offset y at any point, with respect to the cable bottom end, are then given by equations (7.9)–(7.12).

$$H_{\mathrm{b}} = \frac{y_{\mathrm{t}}}{(1/w)\ln(T_{\mathrm{t}}/T_{\mathrm{b}})} \tag{7.8}$$

$$\theta_t = \frac{H_b}{T_t} \tag{7.9}$$

$$\theta_b = \frac{H_b}{T_b} \tag{7.10}$$

$$\text{setdown} = \frac{H_b{}^2 L}{2T_t T_b} \tag{7.11}$$

$$y = \frac{H_b}{w} \ln\left(\frac{T}{T_b}\right) \tag{7.12}$$

Maximum sag (maximum lateral offset relative to the straight line joining cable extremities) occurs at the point where the cable slope is equal to the slope of the line joining the cable extremities (i.e., y_t / L). Hence the tension T_{ms} at the position of maximum sag can be found from:

$$\frac{dy}{dx} = \frac{H_b}{T_{ms}} = \frac{y_t}{L} \tag{7.13}$$

With substitution for H_b from equation (7.8), the tension T_{ms} is given by:

$$T_{ms} = \frac{wL}{\ln(T_t / T_b)} \tag{7.14}$$

Hence, from equation (7.12), the maximum sag, relative to the straight line joining cable extremities, is given by

$$\max \text{sag} = \frac{H_b}{w} \ln\left(\frac{T_{ms}}{T_b}\right) - \left(\frac{T_{ms} - T_b}{wL}\right) y_t \tag{7.15}$$

With substitution for H_b from equation (7.8), equation (7.15) becomes

$$\max \text{sag} = \left[\frac{\ln (T_{ms} / T_b)}{\ln (T_t / T_b)} - \frac{T_{ms} - T_b}{wL}\right] y_t \tag{7.16}$$

Segmented Cable with Current Load

For a near-vertical cable composed of segments with different apparent weights and current loads, the equations are slightly more complicated than those for a uniform cable. By use of the symbols of figure 7–1*b,* equation (7.4) can be applied to a segment to give the lateral offset component Δy_t in terms of H_p, the horizontal component of effective tension at the segment lower end. For segment *pq,* the lateral offset component is given by:

$$\Delta y_t = \frac{H_p}{w_{pq}} \ln\left(\frac{T_q}{T_p}\right) - \int_p^q \frac{\int_0^x f(x)\,dx}{T}\,dx \qquad (7.17)$$

The force H_p at the segment lower end is related to H_b at the cable bottom end by:

$$H_p = H_b - F_{c0p} \qquad (7.18)$$

where F_{c0p} is the total current force acting on the cable between the cable bottom end and the lower end of the segment. Hence, equation (7.17) can be expressed in terms of H_b for each segment. The offset components Δy_t can then be summed for all the segments to give equation (7.19), which allows the horizontal force H_b at the cable bottom end to be found:

$$y_t = H_b \sum\left[\frac{1}{w_{pq}} \ln\left(\frac{T_q}{T_p}\right)\right] - \sum\left[\frac{F_{c0p}}{w_{pq}} \ln\left(\frac{T_q}{T_p}\right)\right] - \sum\left[\int_p^q \frac{\int_0^x f(x)\,dx}{T}\,dx\right] \qquad (7.19)$$

Segmented Cable with Zero Current Load

The resulting equations are again very simple for a segmented cable without current. The top-end offset y_t is the sum of the offset components of all the segments. Use of equation (7.19) with zero current load yields equation (7.20) for the horizontal component of tension:

$$H_b = \frac{y_t}{\sum\left[\left(1/w_{pq}\right)\ln\left(T_q/T_p\right)\right]} \qquad (7.20)$$

The equations for the top- and bottom-end angles and the setdown are then given by equations (7.21)–(7.23):

$$\theta_t = \frac{H_b}{T_t} \tag{7.21}$$

$$\theta_b = \frac{H_b}{T_b} \tag{7.22}$$

$$\text{setdown} = \frac{H_b^{\ 2}}{2} \sum \left(\frac{L_{pq}}{T_p T_q} \right) \tag{7.23}$$

where the parameters w_{pq}, T_p and T_q, and L_{pq}, are the apparent weight per unit length, the effective tensions at the segment extremities, and the length of a segment *pq*, respectively.

The lateral offsets have to be found segment by segment, working up from the bottom end. Within a segment *pq,* the lateral offset *y* of a particular point with respect to the cable bottom end is given by

$$y = y_p + H_b \left[\frac{1}{w_{pq}} \ln \left(\frac{T}{T_p} \right) \right] \tag{7.24}$$

where y_p is the lateral offset of the lower end of the segment, with respect to the cable bottom end, and T is the effective tension at the point concerned.

Simple Approximate Solutions for Near-Vertical Cables

The preceding equations are perfectly exact, within the limits of small-angle deflection theory, and they are simple to program. Nevertheless, the design engineer might appreciate even simpler, *approximate* expressions, which would allow him or her to made quick estimates of the horizontal reaction H_b, the top and bottom angles, the setdown and the profile.

A simple approximation for $\ln(T_t/T_b)$ is given by

$$\ln\left(\frac{T_t}{T_b}\right) \approx \frac{T_t - T_b}{\sqrt{T_t T_b}} \qquad (7.25)$$

The precision of the approximation in equation (7.25) depends on the value of the ratio T_t/T_b (or, since $wL = T_t - T_b$, on the *tension factor* T_t/wL). The precision improves as the tension factor increases. For $T_t/wL = 1.3$, the error is 9%. For $T_t/wL = 1.5$ and 1.7, the errors are 5% and 3%, respectively.

Uniform cable with zero current

By defining the angular offset as $\gamma = y_t \,/\, L$ (see fig. 7–1), the approximation of equation (7.25) can be used in equations (7.8)–(7.12), which then become

$$H_b = \gamma\sqrt{T_t T_b} \qquad (7.26)$$

$$\theta_t = \frac{H_b}{T_t} = \gamma\sqrt{\frac{T_b}{T_t}} \qquad (7.27)$$

$$\theta_b = \frac{H_b}{T_b} = \gamma\sqrt{\frac{T_t}{T_b}} \qquad (7.28)$$

$$\text{setdown} = \frac{H_b^{\,2} L}{2T_t T_b} = \gamma^2 \frac{L}{2} \qquad (7.29)$$

$$y = y_t \frac{x}{L_t}\sqrt{\frac{T_t}{T}} \qquad (7.30)$$

Note the approximate setdown given by equation (7.29) is the setdown of a straight cable without curvature.

Segmented cable with zero current

Equation (7.25) can also be used to give an approximate value of the horizontal force H_b for a segmented riser, which can then be used in equations (7.21)–(7.23) to calculate approximate values for the angles at the extremities and setdown. Substitution into equation (7.20) gives

$$H_b = \frac{y_t}{\sum\left(L_{pq}/\sqrt{T_p T_q}\right)} \tag{7.31}$$

where L_{pq}, T_p, and T_q are the length and the tensions at the extremities of segment pq.

Although equation (7.31) provides a good approximation of H_b, it is scarcely simpler to evaluate than equation (7.20), which is therefore to be preferred, since it is exact.

Summary

Exact analytical expressions have been derived for the profile, angles, setdown, and horizontal reactions of near-vertical cables, which may be uniform or divided into segments with different characteristics. The expressions include the effect of current load. Further, ultrasimple *approximate* expressions have also been derived for cables without current. These can be particularly useful for making quick estimates of behavior of *uniform* cables.

The exact cable equations given in this chapter can now be used when comparing the behavior of near-vertical *risers* with that of near-vertical *cables*. This is the subject of the next chapter.

References

[1]Sparks, C. P. 1980. Mechanical behavior of marine risers—mode of influence of principal parameters. *Journal of Engineering Resources Technology, ASME*. 102 (December), 214–222.

8

Near-Vertical Riser Static Behavior

The object of this chapter is to determine whether the simple expressions derived *analytically* in chapter 6 can be applied to risers that are *not* under constant tension. Such risers have to be analyzed *numerically,* since for the general case, the following basic equation (derived in appendix A; see equation [A.6]) has no analytical solution:

$$EI\frac{d^4y}{dx^4} - T\frac{d^2y}{dx^2} - w\frac{dy}{dx} - f(x) = 0 \qquad (8.1)$$

Note that throughout this chapter, *tension* always implies effective tension.

Riser Linear Behavior

Note that equation (8.1) is linear. Hence, for near-vertical risers, the law of superposition applies, just as it does for near-vertical cables (see chap. 7). For near-vertical risers, the curvature, angles, and lateral displacements resulting from top-end offset, end moments, and current load can therefore all be calculated separately and summed. The effect of the combined loads is equal to the sum of the effects of the separate loads.

Note, however, that superposition does not apply to risers subject to large deflections, such as SCRs, or to risers for which top tension is not constant (i.e., varies with load). Neither does it apply to riser top-end setdown, which is a secondary, nonlinear effect.

Excel Files *Uniform-Riser.xls* and *Segmented-Riser.xls*

This chapter refers to the Excel files *Uniform-Riser.xls* and *Segmented-Riser.xls* provided with the book, in which riser results are compared with cable results for uniform and segmented risers. Riser results take into account the bending stiffness and are calculated *numerically* by dividing the riser into 600 elements. Cable results are obtained analytically by using the equations of chapter 7 with the same data, apart from the bending stiffness, which is put equal to zero. Each Excel file has three work sheets called *Curvature*, *Stiffness,* and *Angles,* which are specifically adapted to test the validity of the conclusions of chapter 6 with respect to curvature, end rotational stiffness, end shear, and end angles deduced from cable end angles.

Curvature and Profiles

In chapter 6, for beams under constant tension, it was shown that beam curvature is close to cable curvature everywhere apart from zones close to the supports. It can be demonstrated that this is also true for risers by studying different examples, even though it cannot be proved analytically, as it was for beams under constant tension. Table 8–1 gives example data used with the *Curvature* work sheet of *Uniform-Riser.xls*.

The current velocity v is assumed to vary linearly between the riser extremities, and the drag coefficient C_D and the hydraulic diameter ϕ are assumed to be constant over the length of the riser. The drag force at any point is then given by $0.5\rho C_D \phi v^2$, where ρ is the mass density of seawater.

Figure 8–1*a* shows the riser and cable curvatures obtained with *Uniform-Riser.xls,* for the data of table 8–1. The curvatures are seen to be in close agreement over the entire length except for zones close to the supports, as expected from the results of chapter 6.

Table 8–1. *Example data for a uniform riser*

Top end offset (m)		45
Length (m)		600
Top tension (kN)		2,000
Total apparent weight (kN)		1,500
Bending stiffness, EI (kN-m^2)		318,600
Moments (kN-m)	Top end	0
	Btm end	0
Drag coefficient, C_D		0.7
Hydraulic diameter, ϕ (m)		1
Current velocities (m/s):	Top end	1
	Btm end	0.1

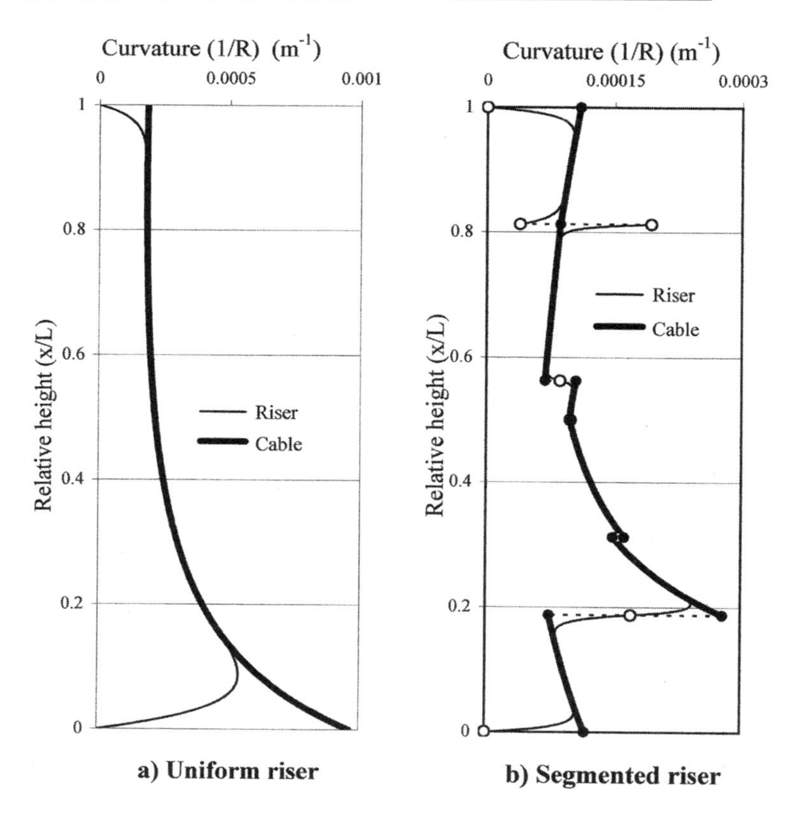

Fig. 8–1. *Riser curvature compared to cable curvature—two examples*

The *Curvature* work sheet of *Uniform-Riser.xls* can provide similar comparisons between riser and cable curvature, using different data. The file also outputs and compares global results, including end angles, maximum sag with its position, top-end setdown, and the horizontal end reactions.

In chapter 6, the extent of the zones of influence of the supports depended on the flexibility factor kL (see equation [6.4]), which is an important parameter in the analytical equations. Although kL cannot have the same significance for a riser with varying tension, it can be used to give general indications about riser flexibility. Indicative values of kL are therefore given by the file, for three values of tension: top tension, mean tension, and bottom-end tension.

The convergence between riser and cable curvature is even more striking for risers made up of segments with different characteristics, as can be demonstrated using the *Curvature* work sheet of *Segmented-Riser.xls*. Table 8–2 gives example data.

Table 8–2. *Example data for a segmented riser*

Top end offset (m)	120				
Top tension (kN)	5,200				
Moment top end (kN-m)	0				
Moment btm end (kN-m)	0		Drag coeff. C_D		0
Data per Segment	*App.wt (kN/m)*	*Length (m)*	*Stiffness EI (kN-m²)*	*Hydraulic diameter φ (m)*	*Cu vel (n*
Segment 1 (top)	1	300	1,593,000	1	1
Segment 2	1	400	318,600	1	
Segment 3	4	100	318,600	1	0
Segment 4	4	300	318,600	1	C
Segment 5	4	200	318,600	0.4	C
Segment 6 (bottom)	1	300	318,600	0.4	C
	Total:	1,600		*Bottom end current :*	C

The characteristics of each segment have to be specified separately, including the current velocity at the top end of each segment. The bottom-end current velocity is input separately, as shown in table 8–2. The current velocity v is assumed to vary linearly over the length of each segment, and the drag coefficient C_D is assumed to be constant over the entire length of the riser. The drag force at any point is then given by $0.5\rho C_D \phi v^2$, where ρ is the mass density of seawater.

Figure 8–1*b* shows the curvature plot for the data of table 8–2, which have been chosen to emphasize tendencies. As can be seen, riser and cable curvatures are still closely matched except for zones near the extremities and for the segment junctions with discontinuities in apparent weight, bending stiffness, or current load (owing to a change in hydraulic diameter).

In Figure 8–1*b,* there is a large discontinuity in the riser curvature at $x/L = 0.81$, resulting from the change in bending stiffness by a factor of five between segments 1 and 2. There is, of course, no discontinuity in *riser moment,* which must always be continuous. At the junction of segments 5 and 6 (at $x/L = 0.19$) and, to a lesser extent, the junction of segments 2 and 3 (at $x/L = 0.56$), there are discontinuities in the cable curvature, resulting from changes in apparent weight per unit length. At the junction of segments 3 and 4 (at $x/L = 0.31$), there is a small discontinuity in cable curvature, resulting from a change in the current load produced by the change in hydraulic diameter.

The *Curvature* work sheet of *Segmented-Riser.xls* can provide comparisons between riser and cable curvature, using different data. The file also outputs and compares global results (end angles, maximum sag with its position, top-end setdown, and horizontal end reactions). Curvatures, moments, and offsets are also tabulated for 25 points—at the junctions between segments and at equally spaced points between the junctions.

The file *Segmented-Riser.xls* can be used to explore other tendencies. For example, if two riser simulations are compared in which the bending stiffness of only one segment is increased by some large factor, then the moments in that section will be increased virtually by the same factor. Curvature will be hardly changed.

Figure 8–2 gives the riser and cable profiles for the data of tables 8–1 and 8–2. At the scale of the figures, no difference can be seen between them. This is not surprising since the offsets obtained with *Segmented-Riser.xls* differ by less than 4 cm.

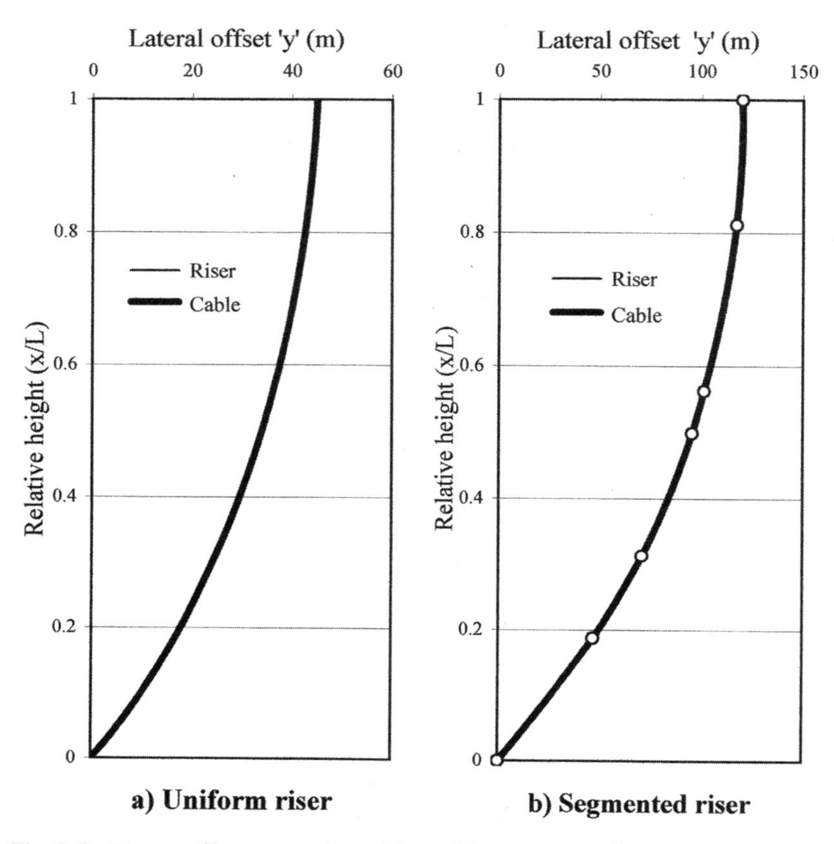

Fig. 8–2. *Riser profile compared to cable profile—two examples*

End Rotational Stiffness

In chapter 6, for beams under constant tension, it was seen that the end rotational stiffness under an applied moment is given closely by $\sqrt{TEI}$ (see equation [6.19]) if the flexibility factor $kL > 10$. That approximation was seen to improve as kL increases. It was also seen that bending is concentrated close to the extremity where the end moment is applied (see the section "Influence of End Moment"). Hence for the bottom end of a riser, the same approximation should be valid, providing that a slightly greater tension is used to take account of the *tension gradient*.

In 1980, I proposed to use a tension corresponding to a point at height $1/k_b$ above the bottom end, where $k_b = \sqrt{T_b / EI}$.[1] This implies using a tension $T = (1 + q_b)T_b$ in equation (6.19), where q_b is a dimensionless parameter defined in equation (8.2) in terms of the apparent weight per unit length w and the bottom-end tension T_b:

$$q_b = \frac{w}{k_b T_b} \qquad (8.2)$$

The *Stiffness* work sheet of *Uniform Riser.xls* allows the end rotational stiffness to be explored for risers. It demonstrates that a more accurate value for the rotational stiffness at the *riser bottom end* is obtained using equation (8.3), which is semiempirical:

$$M_0 \bigg/ \left(\frac{dy}{dx}\right)_0 = (1 + q_b)\sqrt{T_b EI} \qquad (8.3)$$

The upper part of table 8–3 gives an example of bottom-end rotational stiffness as output by the file for a riser with the characteristics given in table 8–1. The exact value of the rotational stiffness (M/θ) is compared with $\sqrt{T_b EI}$ and with the value given by equation (8.3). Since the problem is linear, the values of stiffness are independent of the moment input by the user. As can be seen from the example in table 8–3, the simple formula of equation (6.19) already gives a good value for the rotational stiffness. Nevertheless, equation (8.3) is more accurate.

Table 8–3. *Example results for end rotational stiffness and end shear*

Rotational stiffness		**kN-m/rad**	**Precision ratio**
	Exact value	29,242	-
Bottom end	$\sqrt{(T_b\ EI)}$	27,652	0.94
	$(1+q_b)\sqrt{(T_b\ EI)}$	29,422	1.00
Shear force		**kN**	**Precision ratio**
Bottom end	Exact value	-85.64	-
	-kMo	-86.79	1.01

For the riser top-end stiffness, the reader would probably expect a formula equivalent to equation (8.3) to be applicable, with q_b replaced by $q_t = -w/(k_t T_t)$. However, the *Stiffness* work sheet of *Uniform-Riser.xls* shows that the top-end stiffness is generally given more accurately simply by

$$M_t \bigg/ \left(\frac{dy}{dx}\right)_t = \sqrt{T_t EI} \qquad (8.4)$$

The *Stiffness* work sheet of *Segmented Riser.xls* can be used to verify that equations (8.3) and (8.4) are also generally applicable to segmented risers. The equations for the bottom-end rotational stiffness become less accurate for risers with very low bottom-end tension (i.e., for risers with very low tension factors T_t/wL).

End Shear

When a riser is subject to an end moment, an associated shear force F_0 acts perpendicular to the riser axis. In chapter 6, for a beam under constant tension, the end shear force F_0 was related to the end moment M_0 by equation (8.5), where $k = \sqrt{T / EI}$:

$$F_0 = -kM_0 \qquad (8.5)$$

The *Stiffness* work sheets of the two Excel files also output the *exact shear force* (calculated numerically) associated with a moment at the riser extremity and compare it with the value given by equation (8.5). In equation (8.5), $k = \sqrt{T_t / EI}$ for top-end shear, and $k = \sqrt{T_b / EI}$ for bottom end shear, where T_t and T_b are the top- and bottom-end tensions.

The lower part of table 8–3 gives an example of *bottom-end shear* for the data of table 8–1 with an applied bottom-end moment of 1,000 kN-m. Equation (8.5) is seen to give a very accurate value for the shear force F_0. Note equation (8.5) becomes less accurate as the bottom-end tension approaches zero (i.e., for risers with very low tension factors T_t/wL).

Riser Angles Deduced from Cable Angles

In chapter 6, it was shown that beam end angles can be estimated from cable end angles by taking into account the local straightening at the beam extremities resulting from the bending stiffness. This is also possible for risers.

The curvature $1/R$ is given at the cable top and bottom ends by equations 8.6 and 8.7 respectively:

$$\left(\frac{1}{R}\right)_{\mathrm{t}} = \left(\frac{w\theta_{\mathrm{cable}} + f}{T}\right)_{\mathrm{t}} \tag{8.6}$$

$$\left(\frac{1}{R}\right)_{\mathrm{b}} = \left(\frac{w\theta_{\mathrm{cable}} + f}{T}\right)_{\mathrm{b}} \tag{8.7}$$

where w, θ_{cable}, f and T are the apparent weight per unit length, the end angle, the current load, and the tension, respectively. In equation (8.6) the external subscript "t" implies that top end values have to be used for those four parameters when calculating the top end curvature. In equation (8.7) bottom end values have to be used.

If the moments given by equations (8.8) and (8.9) are applied to the ends of the riser, it should have virtually the same curvature as the cable over its entire length:

$$M_{\mathrm{t}} = \frac{EI_{\mathrm{t}}}{R_{\mathrm{t}}} = (w\theta_{\mathrm{cable}} + f)_{\mathrm{t}}\left(\frac{EI}{T}\right)_{\mathrm{t}} \tag{8.8}$$

$$M_{\mathrm{b}} = \frac{EI_{\mathrm{b}}}{R_{\mathrm{b}}} = (w\theta_{\mathrm{cable}} + f)_{\mathrm{b}}\left(\frac{EI}{T}\right)_{\mathrm{b}} \tag{8.9}$$

If the angular rotational stiffness is known at each end, then the angular effect of releasing the moment can be estimated. Best results are found empirically to be given with an *angular stiffness* of $(1+2q)\sqrt{TEI}$ (rather than with the values given by equations [8.3] and [8.4]), where $q_{\mathrm{b}} = w_{\mathrm{b}}/k_{\mathrm{b}}T_{\mathrm{b}}$ and $q_{\mathrm{t}} = -w_{\mathrm{t}}/k_{\mathrm{t}}T_{\mathrm{t}}$. Hence, the angle *corrections* are given by equations (8.10) and (8.11):

$$\delta\theta_{\mathrm{t}} = \frac{M_{\mathrm{t}}}{(1+2q_t)\sqrt{T_{\mathrm{t}}EI_{\mathrm{t}}}} = \left(\frac{(w\theta_{\mathrm{cable}} + f)EI}{T(1+2q)\sqrt{TEI}}\right)_{\mathrm{t}} \tag{8.10}$$

$$\delta\theta_{\mathrm{b}} = \frac{-M_{\mathrm{b}}}{(1+2q_{\mathrm{b}})\sqrt{T_{\mathrm{b}}EI_{\mathrm{b}}}} = -\left(\frac{(w\theta_{\mathrm{cable}} + f)EI}{T(1+2q)\sqrt{TEI}}\right)_{\mathrm{b}} \tag{8.11}$$

where the external subscripts in the right hand parts of the equations apply to all the parameters within the brackets.

Since $EI / \left[T\sqrt{TEI}\right] = 1/kT$, the deduced riser angles are finally given by

$$\theta_{\text{t riser}} = \theta_{\text{t cable}} + \left(\frac{w\theta_{\text{cable}} + f}{kT(1+2q)} \right)_{\text{t}} \tag{8.12}$$

$$\theta_{\text{b riser}} = \theta_{\text{b cable}} - \left(\frac{w\theta_{\text{cable}} + f}{kT(1+2q)} \right)_{\text{b}} \tag{8.13}$$

These equations are semiempirical. Equation (8.13) already figured in section 7 of my 1980 article.[2] The subject has also been treated by other authors.[3]

Table 8–4. *Example riser angles calculated from cable angles*

End Angles	Riser (numerical calc)	Cable (analytical calc)
Top end	1.00°	0.51°
Bottom end	8.49°	10.95°
Riser angles calculated from cable angles		**Precision ratio**
Top end	0.99°	0.98
Bottom end	8.51°	1.00

The *Angles* work sheets of the two files allow the accuracy of equations (8.12) and (8.13) to be tested. Table 8–4 shows an example of riser and cable angles calculated using *Uniform-Riser.xls,* for the data of table 8–1 but with bending stiffness increased by a factor of 10 (to EI = 3,186,000 kN-m²). The upper part of table 8–4 gives the riser and cable angles, calculated numerically and analytically, respectively. The lower part of table 8–4 gives the riser angles obtained from the cable angles by using equations (8.12) and (8.13) to take into account the effects of the bending stiffness. As can be seen, the corrections are very accurate for the example, even though the angle corrections are very large.

Summary

The static behavior of near-vertical risers has been calculated numerically and compared with the results obtained analytically for near-vertical cables. It has been shown that riser curvature is close to cable curvature over the entire length of a riser, apart from zones close to supports or close to junctions between segments in the case of large discontinuities in bending stiffness or apparent weight.

Semiempirical formulae, similar to those of chapter 6, have been derived for the end rotational stiffness at the extremities of a riser with nonconstant tension. The formula derived in chapter 6, for the end shear associated with an end moment, has been shown to be valid for risers

Simple formulae for the deduction of riser end angles from cable end angles have also been derived and verified. The Excel files *Uniform-Riser.xls* and *Segmented-Riser.xls* have been presented and have been used to test all the derived formulae and deduced tendencies.

The results obtained using the formulae presented in this chapter are particularly important for the design of stress joints. This is the subject of the next chapter.

References

[1]Sparks, C. P. 1980. Mechanical behavior of marine risers—mode of influence of principal parameters. *Journal of Engineering Resources Technology, ASME.* 102 (December), 214–222.

[2]Ibid.

[3]Bernitsas, M. M., and P. Papalambros. 1980. Design optimization of risers under generalized static load. Paper presented at Intermaritec '80, Hamburg.

9

Stress Joint Design

Although risers are flexible structures, they are often rigidly connected to structures that have little or no flexibility. At such junctions, bending stiffness plays a key role in determining the local response. Short lengths have to be designed with carefully chosen bending stiffness, to avoid overstressing the riser locally in bending. These short lengths are generally called *stress joints* (SJs).

Production risers in particular are frequently connected to the seabed by rigid connections. They are also sometime rigidly connected to a buoy or a platform at their top ends. Spar risers are generally equipped with keel joints that are far stiffer than the main length of the riser.

The behavior of an SJ will depend on the way the bending stiffness varies along its length. This has to be chosen by the designer. Hence, the logical way to proceed is to first choose the required response and then design the *bending-stiffness function* in order to produce that response. The following are three possible design choices for the SJ:

- Constant curvature (i.e., circular bending)
- Constant maximum bending stress
- Tapered wall thickness (i.e., linearly varying internal and external diameters)

A special case of the last choice is a constant internal diameter and a tapered external diameter. The implications of these design requirements are examined in the following sections. The Excel file *SJ-Design.xls* allows the reader to verify that the bending-stiffness equations developed in the following sections satisfy the design requirements.

In this chapter, it is assumed that the SJ is at the lower end of the riser, fixed vertically to the seabed. In the case of a top-end SJ, the design principles are the similar. However, if the SJ is connected to a small floating structure, such as a buoy, the SJ stiffness may be sufficient to influence the top-end response, in which case an iterative calculation will be necessary.

Note that throughout this chapter, *tension* always implies effective tension.

SJ Forces and Design Requirements

Before proceeding with the design of an SJ, the designer must first determine the maximum allowable moment M_0 that the *riser* can accept at the junction with the SJ. Plainly, the greater the allowable riser moment, the smaller will be the angle through which the SJ will have to turn.

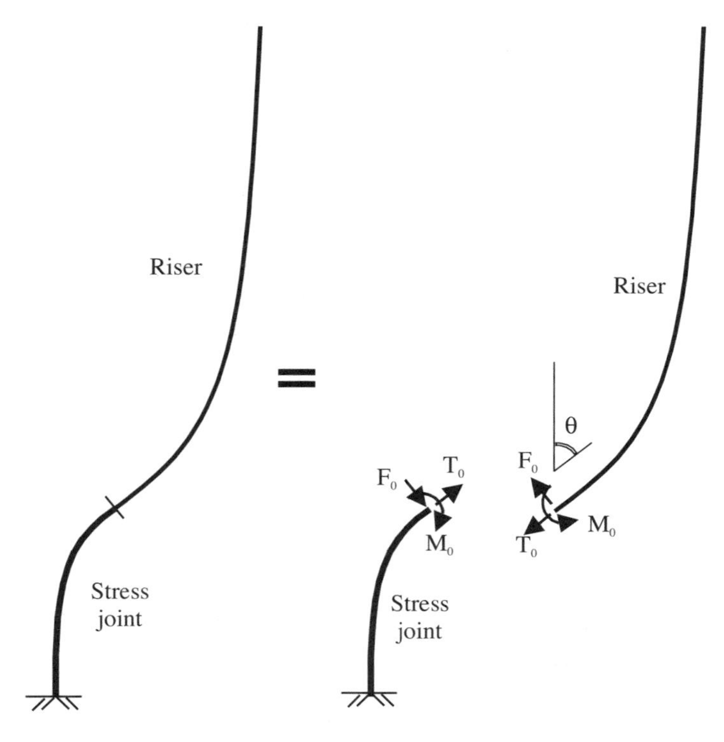

Fig. 9–1. *Forces at junction of SJ and riser*

The riser bottom-end angle θ_b, when moment M_0 is applied (as shown in fig. 9–1), must then be determined. This can be done by using a riser static simulation program or by applying the equations of chapter 8. Thus, the angle θ_j that the SJ must turn through can be determined. For a bottom-end SJ connected vertically to the seabed (as shown in fig. 9–1), angle θ_j is equal to θ_b.

As shown in chapter 8 (see equation [8.5]), the shear force F_0 perpendicular to the riser axis at the riser bottom end is proportional to the riser bottom-end moment M_0 and can be written as follows:

$$F_0 = k_{riser} M_0 \qquad (9.1)$$

where k_{riser} is a constant generally given with good accuracy by $k_{riser} = \sqrt{T_0 / EI_{riser}}$ and T_0 is the tension at the junction of the riser and the SJ. The reader can either accept this value of k_{riser} or use a riser simulation program to determine a more accurate value.

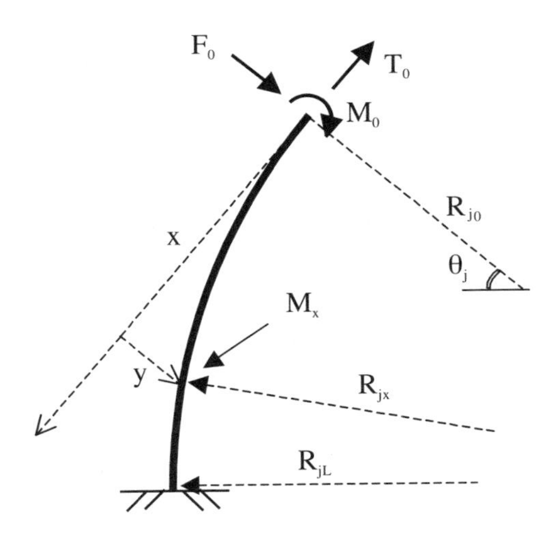

Fig. 9–2. *SJ coordinate system*

Figure 9–2 shows the SJ coordinate system, with the forces applied by the riser to the tip of the SJ. The moment M_0 at the SJ tip will induce a local curvature $1/R_{j0}$ depending on the bending stiffness EI_{j0} of the SJ tip, since $M_0 = EI_{j0}/R_{j0}$. Note that the stiffness EI_{j0} may be different from the riser stiffness EI_{riser}, particularly if the two are made of different materials.

The SJ is considered sufficiently short for the tension T_0 to be constant along its length. By use of the coordinate system of figure 9–2, the moment M_x at distance x from the tip of the SJ is then given by

$$M_x = M_0 + F_0 x + T_0 y \tag{9.2}$$

The designer needs to define the required curvature $1/R_{j\,x}$ as a function of x (see the coordinate system in fig. 9–2). Since the curvature $1/R_{j\,x} = (d^2y/dx^2)_{j\,x}$, the value of y and the slope dy/dx can be found subsequently by integration. Then, the moment function M_x can be determined, and the resulting required bending stiffness function $EI_{j\,x}$ can be deduced.

The designer may specify *constant curvature* for the SJ, which is the simplest case. Generally, an SJ is of tapered conical form with an external diameter ϕ_e that varies linearly with x. Since the bending stress σ_b is related to the bending radius R by $\sigma_b = E\phi_e / 2R$, the designer should more logically design for a bending radius that varies with the outer diameter ϕ_e to induce *constant maximum bending stresses* on the SJ outer surface. These cases are considered in the following sections.

SJ with Constant Curvature

The curvature of the SJ in figure 9–2 is given by $1/R_j = d^2y/dx^2$. If this curvature is chosen to be constant, the deflected shape of the SJ can be found by integration. The moment function and the required bending stiffness function can then be determined.

By use of the coordinate system of figure 9–2, the curvature is given by equation (9.3), and the first and second integrals are given by equations (9.4) and (9.5):

$$\frac{d^2y}{dx^2} = \frac{1}{R_j} = \text{constant} \tag{9.3}$$

$$\frac{dy}{dx} = \left(\frac{1}{R_j}\right)x \tag{9.4}$$

$$y = \left(\frac{1}{R_j}\right)\frac{x^2}{2} \tag{9.5}$$

Since dy/dx is the angle turned through, the required length of the SJ can be deduced immediately from equation (9.4). It is given by

$$L_j = R_j \theta_j \tag{9.6}$$

where θ_j is in radians.

The moment function can then be found by substituting equation (9.5) into equation (9.2), giving

$$M_x = M_0 + F_0 x + T_0 \left(\frac{1}{R_j} \right) \frac{x^2}{2} \tag{9.7}$$

Dividing equation (9.7) by M_0 yields equation (9.8):

$$\frac{M_x}{M_0} = 1 + \left(\frac{F_0}{M_0} \right) x + \left(\frac{T_0}{M_0 R_j} \right) \frac{x^2}{2} \tag{9.8}$$

Since curvature is constant, $1/R_j = M_x / EI_{jx} = M_0 / EI_{j0}$. Hence, $M_x / M_0 = EI_{jx} / EI_{j0}$. By use of equation (9.1) and definition of $k_{j0} = \sqrt{T_0 / EI_{j0}}$, equation (9.8) yields

$$\frac{M_x}{M_0} = \frac{EI_{jx}}{EI_{j0}} = 1 + k_{riser} x + \frac{(k_{j0} x)^2}{2} \tag{9.9}$$

If the bending stiffness of the SJ tip is the same as that of the riser and the value of k_{riser} given following equation (9.1) is accepted, then $k_{riser} = k_{j0}$.

To summarize, the procedure for dimensioning an SJ to obtain *circular bending* is as follows:

- Determine the forces applied by the riser to the SJ tip—namely, the moment M_0, the associated shear force F_0, and the tension T_0.
- Determine the angle θ_j through which the SJ must turn.
- Define the required SJ curvature $1/R_j$ and the corresponding bending stiffness EI_{j0} of the SJ tip, since $1/R_j = M_0 / EI_{j0}$. This should take into account the maximum allowable bending stress given by $\sigma_b = E\phi_e / 2R_j$.
- The required length of the SJ is then given immediately by equation (9.6), and the required bending-stiffness function EI_{jx} is given by equation (9.9).

SJ with Constant Maximum Bending Stress

The procedure for designing an SJ with constant maximum bending stress on the outer surface is similar to that outlined in the preceding section. However, the equations are slightly more complicated.

SJs are generally of tapered conical shape. The maximum bending stress on the outer surface will therefore be constant if the ratio of external diameter to bending radius (ϕ_e/R) is constant.

The designer can choose to impose a ratio α_j between the *bending radii* R_{jL} and R_{j0} at the extremities of the SJ. To obtain constant maximum bending stresses, the same ratio α_j has to be specified for the external diameters ϕ_{eL}/ϕ_{e0} at the extremities. Hence,

$$\alpha_j = \frac{R_{jL}}{R_{j0}} = \frac{\phi_{eL}}{\phi_{e0}} \qquad (9.10)$$

Furthermore, if the external diameter ϕ_{ex} and the bending radius R_{jx} both vary linearly with x, then they will be given by

$$\frac{R_{jx}}{R_{j0}} = \frac{\phi_{ex}}{\phi_{e0}} = 1 + bx \qquad (9.11)$$

where b is a constant (given by equation [9.12]) in terms of α_j and L_j, the length of the SJ:

$$b = \frac{\alpha_j - 1}{L_j} \qquad (9.12)$$

The required curvature function can be deduced from equation (9.11). It is given by

$$\frac{d^2y}{dx^2} = \frac{1}{R_{jx}} = \frac{1}{R_{j0}}\left(\frac{1}{1+bx}\right) \qquad (9.13)$$

The slope and deflection functions can then be found by integration. For the coordinate system of figure 9–2, they are given by equations (9.14) and (9.15):

$$\frac{dy}{dx}=\frac{1}{bR_{j0}}\ln(1+bx) \qquad (9.14)$$

$$y=\frac{1}{b^2R_{j0}}\left[(1+bx)\ln(1+bx)-bx\right] \qquad (9.15)$$

From equation 9.14, the angle turned through by the SJ is $\theta_j=(1/bR_{j0})/\ln(1+bL_j)$, or $\theta_j=(1/bR_{j0})/\ln(\alpha_j)$. Since $b=(\alpha_j-1)/L_j$ from equation 9.12, the required length of the SJ is therefore given by

$$L_j=\frac{\alpha_j-1}{\ln(\alpha_j)}R_{j0}\theta_j \qquad (9.16)$$

where θ_j is in radians. Hence, from equations (9.2) and (9.15), the moment M_x at x is given by

$$M_x=M_0+F_0x+T_0\frac{1}{b^2R_{j0}}\left[(1+bx)\ln(1+bx)-bx\right] \qquad (9.17)$$

Dividing equation (9.17) by M_0 yields

$$\frac{M_x}{M_0}=1+\frac{F_0}{M_0}x+\left(\frac{T_0}{M_0R_{j0}}\right)\frac{1}{b^2}\left[(1+bx)\ln(1+bx)-bx\right] \qquad (9.18)$$

However, $M_0R_{j0}=EI_{j0}$, which is the bending stiffness at the SJ tip. Using equation (9.1) and defining $k_{j0}=\sqrt{T_0/EI_{j0}}$ leads to equation (9.19), which gives the bending moment M_x along the SJ as a function of x:

$$\frac{M_x}{M_0}=1+k_{riser}x+\left(\frac{k_{j0}}{b}\right)^2\left[(1+bx)\ln(1+bx)-bx\right] \qquad (9.19)$$

At all points along the SJ, the bending radius $R_{\mathrm{j}\,x}$, the moment M_x and the bending stiffness $EI_{\mathrm{j}\,x}$ are related by $EI_{\mathrm{j}\,x} = M_x\, R_{\mathrm{j}\,x}$. This, combined with equation (9.11), leads to

$$\frac{EI_{\mathrm{j}x}}{EI_{\mathrm{j}0}} = \frac{M_x}{M_0}\left(\frac{R_{\mathrm{j}x}}{R_{\mathrm{j}0}}\right) = \frac{M_x}{M_0}(1+bx) \qquad (9.20)$$

Hence, from equation (9.19), the required bending-stiffness function $EI_{\mathrm{j}\,x}$ is given by

$$\frac{EI_{\mathrm{j}x}}{EI_{\mathrm{j}0}} = (1+bx)\left\{1+k_{\mathrm{riser}}x+\left(\frac{k_{\mathrm{j}0}}{b}\right)^2\left[(1+bx)\ln(1+bx)-bx\right]\right\} \qquad (9.21)$$

If the bending stiffness of the SJ tip is the same as that of the riser and the value of k_{riser} given following equation (9.1) is accepted, then $k_{\mathrm{riser}} = k_{\mathrm{j}\,0}$.

To summarize, the procedure for dimensioning an SJ to obtain constant maximum bending stresses along its length is as follows:

- Determine the forces applied by the riser to the SJ tip—namely, the moment M_0, the associated shear force F_0, and the tension T_0.
- Determine the angle θ_j to be turned through by the SJ.
- Define the required curvature $1/R_{\mathrm{j}\,0}$ at the SJ tip and corresponding bending stiffness $EI_{\mathrm{j}\,0}$ (since $1/R_{\mathrm{j}\,0} = M_0/EI_{\mathrm{j}\,0}$). This should take into account the maximum allowable bending stress given by $\sigma_\mathrm{b} = E\phi_{\mathrm{e}\,0}/2R_{\mathrm{j}\,0}$. This bending stress will be constant along the SJ.
- Propose a value for α_j that will apply to the ratios of the radii of curvature $R_{\mathrm{j}\,L}/R_{\mathrm{j}\,0}$ and to the ratios of the external diameters $\phi_{\mathrm{e}\,L}/\phi_{\mathrm{e}\,0}$, at the ends of the SJ, as given by equation (9.10).
- Then, the required length of the SJ is given by equation (9.16), and the required bending-stiffness function $EI_{\mathrm{j}\,x}$ is given by equation (9.21).

Equation (9.21) defines the required bending-stiffness function to give constant maximum bending stresses along the SJ. The function is valid for any value of the SJ tip moment M_0 (and its associated shear force F_0) and, hence, for any angle through which the SJ is turned. Note, however, that the function is valid only for one value of tension T_0.

Wall Thickness

Equations (9.16) and (9.21) respectively give the required SJ length L_j and the required *ideal-bending-stiffness function* $EI_{j\,x}$, according to the value of α_j specified by the designer (see equation [9.10]). The wall thickness should then be chosen, as a function of x, to obtain that stiffness function. Note that the resulting wall thickness will not necessarily be constant—or even vary linearly along the SJ.

SJ with Tapered Wall

For simplicity of fabrication, an SJ should preferably have a wall thickness that is uniformly tapered, corresponding to linearly varying external and internal diameters along the length of the SJ. A conical SJ can be defined, with the same length and external dimensions as given previously (see equation [9.11]) and the same wall thicknesses at the extremities, giving the same values of bending stiffness $EI_{j\,0}$ and $EI_{j\,L}$. However, if a *tapered wall thickness* is chosen between those extremities, the resulting bending stiffness function $EI_{j\,x}$ will not correspond exactly to the ideal-stiffness function given by equation (9.21). The angle turned through by the SJ will then be different from that given by equation (9.16), and the radius of curvature R will not vary linearly along the length of the SJ.

The differences between results for an SJ with ideal stiffness as given by equation (9.21) and results for an SJ with such a tapered wall may be small. They can be checked using a numerical program such as the Excel file *SJ-Design.xls*.

Simulation and Verification Using Excel File *SJ-Design.xls*

The preceding equations are simple to program. They have been programmed in the Excel file *SJ-Design.xls,* which can be used with different data to calculate the required length L_j of an SJ, the bending-stiffness functions $EI_{j\,x}$, and the angles turned through for both an SJ with ideal stiffness and one with a tapered wall. Altogether, 11 data items have to be input, grouped as follows:

- The characteristics at the riser bottom end: external diameter, wall thickness, and modulus. These are used to calculate the riser bending stiffness EI_{riser} (and, subsequently, the riser bottom-end bending stresses).
- The tension T_0, the maximum allowable moment M_0, and the corresponding shear force F_0 acting at the riser lower end. Note that the file proposes a value for the shear force F_0 based on equation (9.1), with $k_{riser} = \sqrt{T_0 / EI_{riser}}$.
- The angle θ_j to be turned through by the SJ.
- A value for the ratio α_j of the bending radii (and external diameters) between the ends of the SJ (see equation [9.10]).
- The SJ material modulus E, the SJ tip external diameter $\phi_{e\,0}$, and the wall thickness. These are used to determine the SJ tip bending stiffness $EI_{j\,0}$ and the bending radius $R_{j\,0}$ induced by moment M_0.

SJ-Design.xls calculates the SJ length from equation (9.16). The bending-stiffness function is plotted and tabulated for both SJs, with ideal stiffness and with a tapered wall.

SJ-Design.xls carries out a numerical simulation (using 600 elements), for the two SJs. Riser loads T_0, M_0, and F_0 are applied to the SJ tips. The resulting angles turned through are calculated. The *bending-radius functions* are tabulated and plotted.

Note that for the SJ with *ideal stiffness*, calculated according to the preceding equations, the following results should always be verified:

- The plot of bending-radius function should always be a straight line, since a linear variation (between $R_{j\,0}$ and $\alpha_j R_{j\,0}$) over the length of the SJ was assumed when deriving the equations.
- The calculated bending radii R should correspond exactly to the design objectives.
- The angle turned through should be equal to the required (data) value.
- If data values of M_0, F_0, and θ_j are all changed in the same proportion (for all other data unchanged), the SJ design is not altered (length, end diameters, thicknesses, and stiffnesses remain unchanged). Bending stresses, moments, and curvatures ($1/R$) are all changed in the same proportion.

Note that according to the value of α_j, the plot of the bending radii along the *tapered-wall* SJ may depart considerably from a straight line, in which case the angle turned through will also be different from the required value. Usually, different interesting values of α_j can be quickly found by trial and error, as explained in the example given in the following section.

Note also that if the SJ tip moment is very small, there may be no solution for the value of α_j specified. In that case, there is no output. The value of α_j should be increased.

Numerical Example

SJ-Design.xls is preloaded with an example in which a steel riser is connected to a titanium SJ which is required to turn through 10°. Different values of α_j can be tried. The following are some interesting values:

- α_j = 1.7 gives a good fit between the ideal-stiffness SJ and the tapered-wall SJ. The SJ length is 13.15 m. The angle turned through by the SJ with tapered wall is 9.7° instead of 10°.
- α_j = 2.15 gives a constant wall thickness (0.0127 m), but the fit between the two SJs is less good. The SJ length is increased to 14.97 m. The angle turned through by the tapered-wall SJ is reduced to 9.4°.
- α_j = 1.235 gives a constant internal diameter (0.2191 m). However, the agreement between the plots for the ideal-stiffness SJ and the tapered-wall SJ is much less good. The SJ length is reduced to 11.09 m, but the angle turned through by the the tapered-wall SJ is reduced to 8.9°. The tapered-wall SJ is significantly stiffer than for the ideal-stiffness SJ in the middle region, where the bending radius is 75 m instead of 64 m, which explains why the angle turned through is reduced.
- The angle turned through by the tapered-wall SJ can be increased by modifying the input data value of θ_j. Increasing the required angle to be turned through to θ_j = 10.95° and using α_j = 1.292 again gives a constant internal diameter (0.2191 m), and the angle turned through by the tapered-wall SJ is increased to 10°. The length of the SJ is then increased to 13.12 m.

Summary

In this chapter, the forces that have to be taken into account when designing an SJ have been defined. Simple equations have then been derived that allow an SJ to be dimensioned to respect different laws of bending, as defined by the designer. The simplest set of equations gives the required bending-stiffness function that will lead to circular bending of the SJ.

A second set of equations gives the required bending-stiffness function that will lead to constant maximum bending stress on the external surface of a conical SJ. However, this ideal-stiffness function will not necessarily lead to a tapered wall thickness, which is preferred for simplicity of fabrication.

The Excel file *SJ-Design.xls* has been presented. It can be used to calculate the required length and ideal-stiffness function to give constant bending stress, for different data. The file also carries out numerical recalculations of the responses of two SJs, one with ideal stiffness and the other with a tapered wall.

A numerical example has been used to show how the characteristics of an SJ can be adjusted by trial and error to obtain particular tapered-wall solutions, such as constant wall thickness or constant internal diameter.

10

Riser Bundles: Local Bending Between Guides

It has been shown in chapter 8 that the bending stiffness EI has little influence on riser curvature except for zones close to the extremities. This is also true for the global behavior of a *riser bundle* composed of several pipes linked together by guides. Nevertheless, in a fine analysis, it is necessary to take into account the *local bending* of the individual pipes between the guides. In certain cases, such local bending may be significantly different from the global bending of the bundle. The share of the moments, between the different members of the bundle, may vary significantly according to the load type (apparent weight, hydrodynamic loads, and inertia forces).

Since it is complicated to carry out simulations in which all pipes in a bundle are modeled, the object of this chapter is to explain how the moments in individual pipes can be deduced from the results of a global analysis, by taking into account local bending at and between the guides.

Note that the terms *tension* and *compression* in this chapter always refer to effective tension and effective compression.

General Bundle Behavior

If the guides are sufficiently closely spaced, the global behavior of a riser bundle can be calculated by lumping together the characteristics of all the pipes that make up the bundle: the tensions, the bending stiffnesses, and the lateral loads. However, between the guides, the profile—and, hence, the curvature—of an individual pipe may depart significantly from the mean value of the bundle.

Figure 10–1 illustrates how individual pipes in a riser bundle may deform between guides for three adjacent spans of equal length L_g. The dotted line shows the mean profile of the bundle. According to its particular characteristics (lateral load, tension, and bending stiffness), an individual pipe may deflect between guides more or less than the mean deflection of the bundle, as shown by the thick and thin solid lines in figure 10–1.

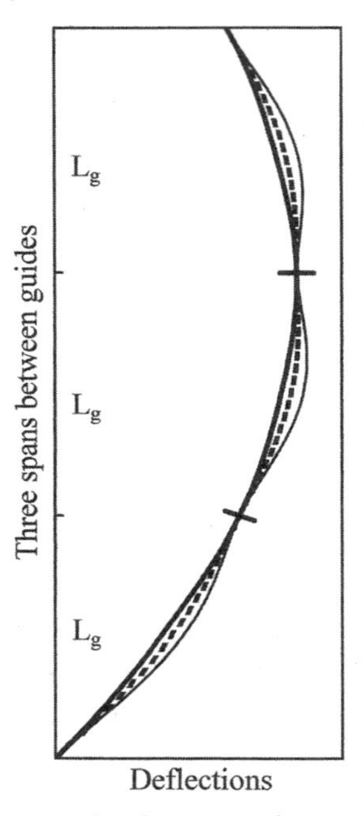

Fig. 10–1 *Riser bundle deflections for three spans between guides*

There will be no discontinuity in the curvature at the guides because of the bending stiffness of the individual pipes. The guides are generally sufficiently closely spaced for the lateral load and the axial tension to be considered constant over several spans. By symmetry at the guides, the profile of all pipes will be asymptotic with the bundle profile, as shown in figure 10–1. The equations for the pipe profiles and moments are derived in appendix G.

M_{pbc} is defined as the moment in a particular pipe that would be induced by bundle curvature ($1/R_{bundle}$). It is the moment that would be induced in the pipe if the bundle moment M_{bundle} were distributed between pipes in proportion to their bending stiffnesses. It is related to the bundle moment and curvature by $M_{pbc} / EI_{pipe} = M_{bundle} / EI_{bundle} = 1/R_{bundle}$. Hence, M_{pbc} is given by

$$M_{\text{pbc}} = \frac{EI_{\text{pipe}}}{R_{\text{bundle}}} = \left(\frac{EI_{\text{pipe}}}{EI_{\text{bundle}}}\right) M_{\text{bundle}} \tag{10.1}$$

In a span between guides, away from the riser extremities, the moment function M_x for the pipe under *tension* is given by the following equation (which is derived in appendix G; see equation [G.9]):

$$\frac{M_x}{M_{\text{pbc}}} = \left[1 - \left(\frac{f}{T}\right) R_{\text{bundle}}\right] \left(\frac{kL_g}{2}\right) \frac{\cosh k\left[\left(L_g/2\right) - x\right]}{\sinh\left(kL_g/2\right)} + \left(\frac{f}{T}\right) R_{\text{bundle}} \tag{10.2}$$

where x is measured from a guide, L_g is the length between guides, f and T are the respective pipe lateral load per unit length and the tension, R_{bundle} is the bundle mean radius of curvature, and $k = \sqrt{T / EI_{\text{pipe}}}$.

For the pipe under *compression,* the moment equation is given by the following equation (see equation [G.18]):

$$\frac{M_x}{M_{\text{pbc}}} = \left[1 + \left(\frac{f}{F}\right) R_{\text{bundle}}\right] \left(\frac{kL_g}{2}\right) \frac{\cos k\left[\left(L_g/2\right) - x\right]}{\sin\left(kL_g/2\right)} - \left(\frac{f}{F}\right) R_{\text{bundle}} \tag{10.3}$$

where F is the pipe compression and $k = \sqrt{F / EI_{\text{pipe}}}$; other symbols are as defined for equation (10.2).

It was shown in chapters 6 and 8 that riser curvature is very close to cable curvature everywhere except for zones close to the supports. Hence, away from the riser extremities, the bundle curvature is given closely by

$$\frac{1}{R_{\text{bundle}}} = \left(\frac{f}{T}\right)_{\text{bundle}} \tag{10.4}$$

The ratio α is defined as follows, noting first that α is negative for the pipe in compression ($F = -T$):

$$\alpha = R_{\text{bundle}} \left(\frac{f}{T}\right)_{\text{pipe}} \tag{10.5}$$

Hence, from equation (10.4),

$$\alpha = \left(\frac{f}{T}\right)_{\text{pipe}} \Big/ \left(\frac{f}{T}\right)_{\text{bundle}} \tag{10.6}$$

or

$$\alpha = \frac{f_{\text{pipe}}}{f_{\text{bundle}}} \bigg/ \frac{T_{\text{pipe}}}{T_{\text{bundle}}} \tag{10.7}$$

Also, β is defined in accordance with the following equation:

$$\beta = \frac{L_g}{2}\sqrt{\frac{|T|}{EI}} \tag{10.8}$$

where $|T|$ is the absolute value of the axial force (tensile or compressive). Then, for a pipe under *tension,* equation (10.2) can be rewritten as follows:

$$\frac{M_x}{M_{\text{pbc}}} = (1-\alpha)\beta\left\{\frac{\cosh\beta\left[1-(2x/L_g)\right]}{\sinh\beta}\right\} + \alpha \tag{10.9}$$

Further, for a pipe under *compression,* equation (10.3) can be rewritten as follows:

$$\frac{M_x}{M_{\text{pbc}}} = (1-\alpha)\beta\left\{\frac{\cos\beta\left[1-(2x/L_g)\right]}{\sin\beta}\right\} + \alpha \tag{10.10}$$

The ratio M_x/M_{pbc} is an *amplification function* to be applied to the moment induced by bundle curvature, given by equation (10.1). M_x/M_{pbc} depends on α, which depends on the *load ratio* $f_{\text{pipe}}/f_{\text{bundle}}$ (see equation [10.7]), which may be different for different load types.

If the load ratio $f_{\text{pipe}}/f_{\text{bundle}}$ is equal to the *tension ratio* $T_{\text{pipe}}/T_{\text{bundle}}$, then $\alpha = 1$ (see equation [10.7]), and $M_x/M_{\text{pbc}} = 1$ (see equations [10.9] and [10.10]). The pipe will then be subject to exactly the same curvature as the bundle.

Note that equations (10.9) and (10.10) cannot be used when the axial force is zero. Both equations then converge to the nontensioned-beam equation given in appendix G (see equation [G.25]). This, when combined with equation (10.4), yields

$$\frac{M_x}{M_{\text{pbc}}} = 1 - \left(\frac{f_{\text{pipe}}}{f_{\text{bundle}}}\right)\frac{T_{\text{bundle}}L_g^2}{12EI_{\text{pipe}}}\left[1 - 6\frac{x}{L_g}\left(1 - \frac{x}{L_g}\right)\right] \tag{10.11}$$

Distribution of Moments According to Load Type

For any particular pipe in a riser bundle, the load ratio f_{pipe}/f_{bundle} will vary for different load types. Hence, the value of α—and, therefore, the amplification function M_x/M_{pbc}—will also depend on the load type. In particular, three load types need to be considered:

- Apparent weight loads (components acting normal to the pipe axis)
- Hydrodynamic loads
- Inertia forces

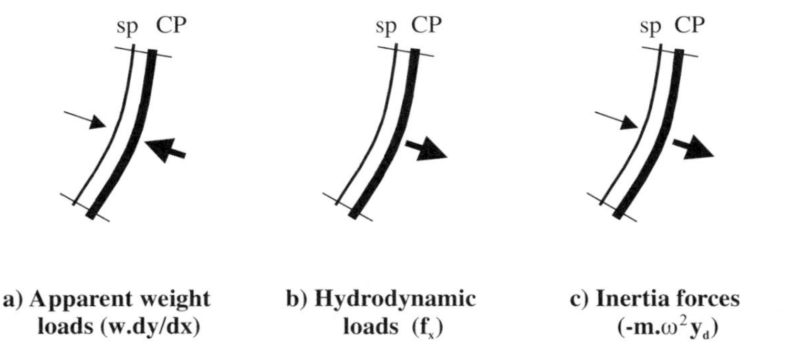

Fig. 10–2. *Example loads on pipes in a riser bundle*

Figure 10–2 shows graphically, for a particular example, how the magnitudes and directions of loads on individual pipes in a riser bundle may be very different for the three load types. The sketches in the figure show just one *satellite pipe* ("sp" in fig. 10–2) and the *core pipe* (CP) of a riser bundle. For the example in figure 10–2, the CP is considered to be equipped with large buoyancy modules that house the satellite pipes and protect them from hydrodynamic loads. The sketches of figure 10–2 are discussed in the following subsections, for each load type.

Apparent weight loads normal to pipe axis

For any particular pipe in a riser bundle, the evaluation of the load ratio f_{pipe}/f_{bundle} owing to apparent weight is straightforward. The load normal to the pipe axis is equal to the pipe apparent weight times the angle with the vertical ($w\ dy/dx$). The mean pipe angle dy/dx is the same as the mean bundle angle; therefore,

$$\frac{f_{\text{pipe}}}{f_{\text{bundle}}} = \frac{w_{\text{pipe}}}{w_{\text{bundle}}} \tag{10.12}$$

where w_{pipe} and w_{bundle} are the apparent weights per unit length of the pipe and bundle, respectively. Hence, from equation (10.7),

$$\alpha = \frac{w_{\text{pipe}}}{w_{\text{bundle}}} \bigg/ \frac{T_{\text{pipe}}}{T_{\text{bundle}}} \tag{10.13}$$

The apparent weight of pipes fitted with buoyancy units may be positive, zero or negative. Hence, for pipes in the same bundle, α may be positive, zero or negative. Figure 10–2*a* shows graphically the apparent weight loads for a case in which a specific satellite pipe has *positive* apparent weight and the CP has large *negative* apparent weight.

Hydrodynamic loads

For hydrodynamic loads, the load ratio $f_{\text{pipe}}/f_{\text{bundle}}$ may vary from 0 to 1, according to the details of the bundle and the particular pipe being considered. According to that ratio, the value of α can then be determined from equation (10.7). If the bundle includes a CP equipped with buoyancy units that protect all the satellite pipes from such loads, as assumed in figure 10–2*b,* then the hydrodynamic load will be taken entirely by the CP. Hence, $\alpha = 0$ for the satellite pipes, and $\alpha = 1$ for the CP. For a bundle without buoyancy units, the estimation of hydrodynamic loads on individual pipes in a riser bundle is difficult and cannot be made very precisely.

Inertia forces

When a pipe vibrates laterally at frequency ω with amplitude y_d, it is subject to a dynamic inertia force $-my_d\omega^2$, where m is the *mass plus added mass* per unit length. For a bundle, if it can be assumed that the pipe vibrations between guides are small compared to the bundle vibration, the frequency ω and the amplitude y_d at any point will be the same for the individual pipes as for the bundle. Hence, the load share will be equal to the mass ratio:

$$\frac{f_{\text{pipe}}}{f_{\text{bundle}}} = \frac{m_{\text{pipe}}}{m_{\text{bundle}}} \qquad (10.14)$$

where m_{pipe} and m_{bundle} are the mass plus added mass of the pipe and bundle, respectively. Note that the mass of pipe contents and of any attached buoyancy modules must also be included in m_{pipe} and m_{bundle}.

Figure 10–2*c* shows graphically the inertia forces acting on a satellite pipe and on the CP of a bundle. They are always in the same direction and can never be zero.

Numerical Application Using Excel File *Bundle-Moments.xls*

The reader can use the Excel file *Bundle-Moments.xls* to calculate the amplification functions M_x/M_{pbc} for different riser bundles. The data format is shown in table 10–1. The factors M_x/M_{pbc} can be calculated for three types of pipe simultaneously. The pipes are called CP, tube A, and tube B. It is not necessary to specify the number of individual pipes in the bundle, which may include further pipes with different characteristics. The values input for the riser bundle must take into account the total number of pipes.

Table 10–1. *Data input for Excel file* Bundle-Moments.xls

Riser total length, L (m)	1,200			
Length between guides, L_g (m)	10			
Height, z (m)	500			
Data per pipe	**CP**	**Tube A**	**Tube B**	**Bundle**
Top tension (kN)	-1,500	360	900	7,500
Apparent weight, w (kN/m)	-3.5	0.3	0.75	4.4
Hydrodynamic load proportion	1	0	0	1
Mass plus added mass (tonnes/m)	2.5	0.05	0.1	4
Bending stiffness, EI (kN-m^2)	180,000	2,400	20,000	400,000
Tension at height z, T_z (kN)	950	150	375	4,420

Since the amplification functions M_x/M_{pbc} vary with the tension, they will have different values at different positions along the riser. For convenience, the top tension is input individually for each of the three pipes (CP, tube A, and tube B), as well as for the bundle. The tensions at a point of interest, at height z above the riser bottom end, are calculated from the apparent weights per unit length and are shown in the bottom line of table 10–1.

The amplification functions M_x/M_{pbc} are calculated using equation (10.9) or (10.10), according to whether the pipe is in tension or compression. The results are both plotted and tabulated.

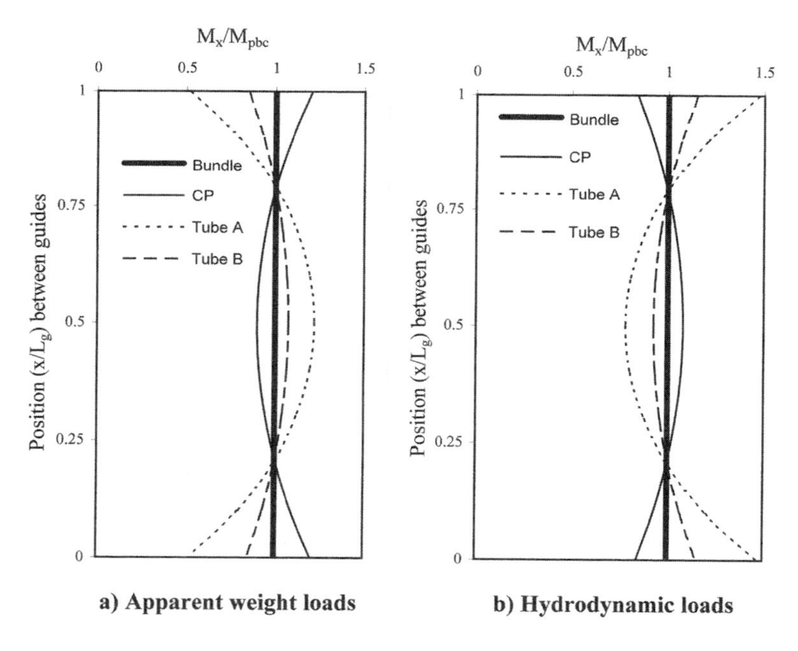

Fig. 10–3. *Example plots of amplification functions M_x/M_{pbc}*

Figure 10–3 shows example plots of M_x/M_{pbc} obtained with the data of table 10–1 for apparent weight loads normal to the pipe axis and for hydrodynamic loads. As can be seen, they are very different. At midspan, for apparent weight loads, M_x/M_{pbc} is greater than 1 for tubes A and B but less than 1 for the CP. This follows because the ratio f/T for the bundle is less than for tubes A and B but greater than for the CP.

For the hydrodynamic loads, the effects are opposite. The tension in the tubes has the effect of trying to straighten tubes A and B. This reduces the curvature and hence reduces the moments at midspan, between guides. However, it increases curvature and moments at the guides.

The plots of M_x/M_{pbc} for the inertia loads for this example are not shown on figure 10–3. They are similar to the plots for the hydrodynamic loads, but with smaller effects.

Decomposition and Recomposition of Moments

As has been shown previously, when calculating the local moments in a pipe between guides, different amplification functions have to be applied to the moment components induced in the pipe by bundle curvature, according to the load type. This may sound complicated, since it implies that the bundle curvature has to be decomposed into its components induced by the different load types. Fortunately, this is not difficult.

It was shown in chapter 8 that the behavior of a near-vertical riser is *linear*. For a near-vertical riser bundle, such as a riser tower, it is therefore possible to obtain the three components of bundle curvature by carrying out the following separate simulations:

- Riser bundle with *top-end offset only,* to obtain the bundle curvature due to apparent weight
- Riser bundle, with zero top-end offset, subject to *current load,* to obtain the bundle curvature due to hydrodynamic loading
- Riser bundle, with zero top-end offset, subject to *lateral vibration,* to obtain the bundle curvature due to inertia loads

For each pipe in the bundle, the amplification functions can then be applied to the different moment components, at the guides and between them. The amplified moment components can then be summed to obtain the total moment at the guides and at points between them.

Summary

This chapter has examined how moments in individual pipes of a riser bundle vary between guides. Amplification functions have been derived that must be applied to the moments induced by the riser bundle curvature obtained from the riser global analysis.

It has been shown that the bundle moments are not distributed between the pipes in simple proportion to bending stiffness. Rather, the moment shared between pipes depends on the distribution of tension and the distribution of load between the individual pipes that make up the bundle.

The moment amplification functions can be very different, according to the load type. Three load types have been considered—namely, apparent weight loads normal to the pipe axis, hydrodynamic loads, and inertia forces.

The linear behavior of near-vertical riser bundles allows the bundle curvature to be decomposed into its components induced by the three load types. Then, the corresponding amplification functions can be applied to the pipe moments induced by bundle curvature, and the resultant amplified components can be re-summed to give the local pipe moments between guides for individual pipes.

The Excel file *Bundle-Moment.xls* allows the reader to explore the amplification functions to be applied at different points along individual pipes of a riser bundle, using different data.

11

Near-Vertical Risers Associated with Floating Platforms with Stiff Tensioners

Near-vertical risers are frequently referred to as top-tensioned risers (TTRs), even though all risers require top tension. Such top tension is often applied using tensioners with significant stiffness. This is the case for TLP production risers in particular.[1] Spar risers are generally tensioned using buoyancy cans, but at the time of this writing, there is already one spar that is equipped with tensioners.[2] There are also several other projects for using tensioners on spars and other floating platforms, for benign environments.[3] Further projects are under study for making risers from new materials, such as high-performance composites, that might allow risers to be *locked off* directly to a TLP or a floating platform, which is equivalent to using infinitely stiff tensioners.[4]

This chapter examines the behavior of near-vertical risers with stiff tensioners associated first with TLPs and then with floating platforms. Simple expressions for the influence of top-end offset on the stability, tension, profile, and sag of such risers are derived. These expressions are based on results of chapters 7 and 8. Chapter 8 showed that bending stiffness has little influence on the profile and setdown of near-vertical deepwater risers. Simple cable expressions for the profiles, sag, and setdown of risers without bending stiffness were derived in chapter 7.

The Excel files *TLP-Risers.xls* and *Floater-Risers.xls* allow the reader to explore the behavior of risers associated with different platforms, using different data. The influence that riser internal changes of temperature, pressure, and fluid densities have is then examined using the results obtained in chapters 3 and 5.

The case of *composite risers,* with their special characteristics, is then considered. The particular problems generated when steel *tubings* are associated with such risers are explained.

Note that throughout this chapter, *tension* always implies effective tension, except for the sub-section on *balanced expansion joints*, where the meanings are explained.

TLP Riser Stretch and Setdown Due to Platform Offset

This section explains how TLP offset y_t, with its associated setdown, influences riser stretch—and, hence, riser tension and sag. This is done for risers fixed to the platform at the top end, with or without tensioners.

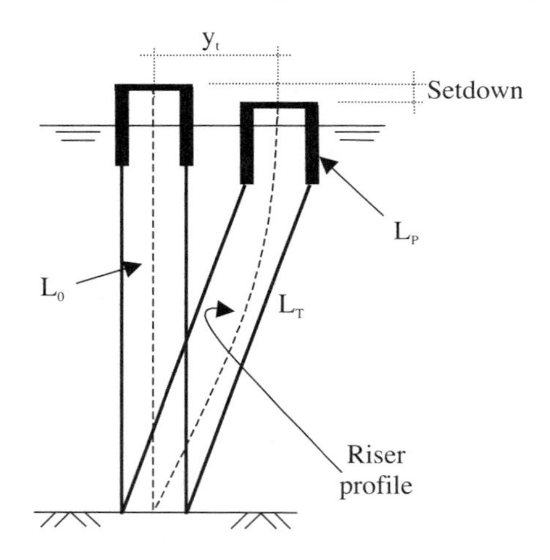

Fig. 11–1. *TLP offset and setdown*

Figure 11–1 shows schematically the TLP before and after the top end is offset laterally by y_t. If the riser is locked off to the TLP deck, the riser top end is plainly subject to the same lateral and vertical movements as the platform. The riser top tension will be modified by the top-end movement, which will cause the riser to stretch (or shrink) axially.

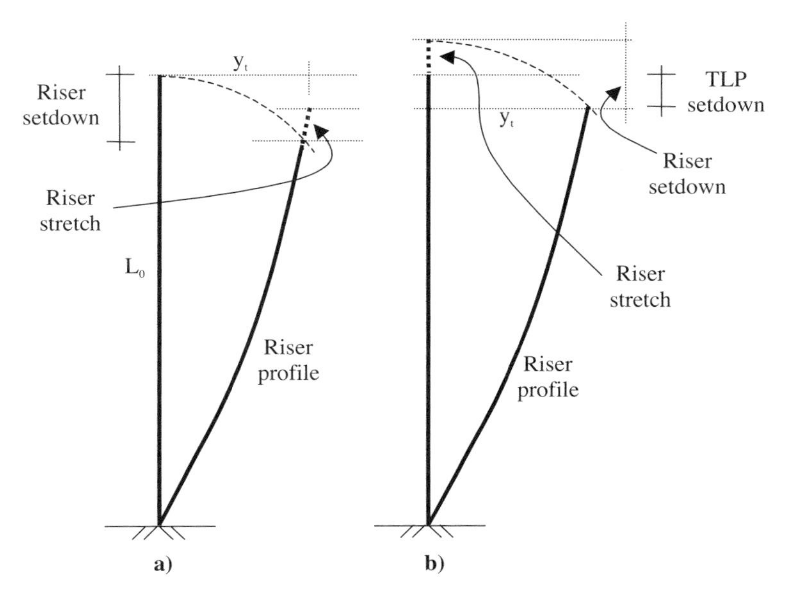

Fig. 11–2. *Relationship between riser stretch and setdown*

The effect of TLP offset and setdown on the riser may be thought of in two ways. As shown in figure 11–2*a,* the riser top end can be thought of as first being offset and setdown, under constant initial top tension $T_{t\ initial}$. Note that the riser setdown under constant tension will be different from the TLP setdown because of the riser sag and because the riser length is greater than TLP tendon length. The riser top end can then be thought of as being displaced vertically to bring it to the TLP deck level, which will stretch (or shrink) the riser, changing the riser tension and sag. The final top tension $T_{t\ final}$ in the offset position has to be found by iteration.

It is probably clearer to consider the procedure shown schematically in figure 11–2*b,* which leads to the same result. The riser can be thought of as being first stretched vertically, by increasing the top tension to $T_{t\ final}$, and then offset laterally by y_t under constant top tension, which will cause the riser top end to set down to the precisely the TLP deck level. The final top tension $T_{t\ final}$ still has to be found by iteration, but the equation relating stretch to setdown is clearly given by

$$\text{riser setdown} - \text{riser stretch} = \text{platform setdown} \qquad (11.1)$$

where the riser setdown is calculated using the final top tension $T_{t\ final}$.

If the *riser excess setdown* is considered to be the riser setdown, under constant final top tension $T_{t\ final}$, less the platform setdown, then from equation (11.1),

$$\text{riser excess setdown} = \text{riser stretch} \qquad (11.2)$$

For the case of a riser with tensioners, the riser stretch is the stretch of the combined system *riser plus tensioners*.

Riser stretch

The platform offset y_t will cause the riser top tension to increase from $T_{t\ initial}$ to $T_{t\ final}$. For the general case, with tensioners, the stretch induced by a tension increase to T_t is given by

$$\text{riser stretch} = \left(T_t - T_{t\ initial}\right)\frac{1}{k_{r+t}} \qquad (11.3)$$

where k_{r+t} is the axial stiffness of the riser plus tensioners, as given in chapter 5 by equation (5.44). Equation (11.3) can be rewritten in terms of the tension factors (T_t/wL_0) as

$$\text{riser stretch} = \left(\frac{T_t}{wL_0} - \frac{T_{t\,initial}}{wL_0}\right)wL_0\frac{1}{k_{r+t}} \qquad (11.4)$$

Note that for locked-off risers without tensioners, the axial stiffness in equation (11.4) is equal to the riser stiffness ($k_{r+t} = k_{riser}$).

Influence of third-order effects

The evolution of riser tension with TLP offset depends on the platform setdown, which is a *second-order* effect induced by offset. For a final riser design, a number of *third-order* effects have to be taken into account. These include the following:

- Tendon extension due to change in platform upthrust resulting from platform setdown, tide, and modified riser tensions. Tendon extension slightly *reduces* platform setdown.

- Tendon sag due to non-zero tendon apparent weight. Such sag *increases* platform setdown. Generally, though, tendons are subject to very high tension and have little sag. Furthermore, for depths down to around 1,000 m, they are often designed to be air filled and virtually neutrally buoyant. For greater depths, in the case of steel tendons, consideration of collapse leads to tendons that are increasingly heavy in water and, hence, subject to some sag.
- Current loads on both the riser and the tendons. These *increase* the calculated setdown of both the riser and the platform.

TLP Riser Tension and Sag Due to Offset: A Simplified Calculation

The Excel files introduced in chapter 8 have shown that the bending stiffness EI generally has very little influence on the profile and the setdown of a deepwater riser. Hence, if the third-order effects mentioned in the previous section are neglected, as well as the riser bending stiffness, the cable equations of chapter 7 can be used to obtain good approximate values for the final riser top tension $T_{\mathrm{t\,final}}$ and the change in riser profile induced by platform offset. The change in profile is characterized by the maximum sag, which is the maximum distance of the riser axis from a straight line joining the riser extremities. Such calculations can be helpful in understanding riser behavior.

For simplified calculations, the tendons are considered to be straight and inextensible. From figure 11–1, the initial vertical riser length L_0 is then related to the tendon length L_{T} and platform column height L_{P} by

$$L_0 = L_{\mathrm{T}} + L_{\mathrm{P}} \tag{11.5}$$

From small-angle deflection theory, the platform setdown is then given as follows:

$$\text{platform setdown} = \frac{y_{\mathrm{t}}^2}{2L_{\mathrm{T}}} \tag{11.6}$$

The offset-induced *riser setdown* under constant tension is as given in chapter 7 by equations (7.8) and (7.11), which combine to give

$$\text{riser setdown} = \frac{L_0}{2T_t T_b\left[(1/w)\ln(T_t/T_b)\right]^2} y_t^{\,2} \qquad (11.7)$$

where T_t and T_b are the respective riser top- and bottom-end tensions and w is the riser apparent weight per unit length.

The *riser excess setdown* as defined preceding equation (11.2) is equal to riser setdown (given by equation [11.7]) minus the platform setdown (given by equation [11.6]). Taking the difference and rearranging gives

$$\text{riser excess setdown} = \left\{\frac{1}{(T_t/wL_0)(T_b/wL_0)\left[\ln(T_t/T_b)\right]^2} - \frac{L_0}{L_T}\right\}\frac{y_t^2}{2L_0} \qquad (11.8)$$

However, from equation (11.2), the riser excess setdown given by equation (11.8) is equal to the riser stretch given by equation (11.4). The value of the final top tension $T_{t\,\text{final}}$ that satisfies that equality has to be found by iteration.

The simplest procedure is to plot the riser excess setdown and the riser stretch given by equations (11.8) and (11.4), respectively, as a function of the tension factor T_t/wL_0, (noting that $T_b=T_t-wL_0$). The final tension factor $T_{t\,\text{final}}/wL_0$ is then given by the intersection of the two curves.

Once the final tension factor has been determined, the position and value of the maximum sag can be found using the equations of chapter 7. From equation (7.14), the tension T_{ms} at the point of maximum sag is given by

$$T_{ms} = \frac{wL_0}{\ln\left(T_{t\,\text{final}}/T_{b\,\text{final}}\right)} \qquad (11.9)$$

The maximum sag therefore occurs at height L_{ms}:

$$L_{ms} = \left(\frac{T_{ms} - T_{b\,\text{final}}}{wL_0}\right)L_0 \qquad (11.10)$$

From equation (7.16), the maximum sag is then given by

$$\text{max sag} = \left[\frac{\ln\left(T_{\text{ms}} / T_{\text{b final}}\right)}{\ln\left(T_{\text{t final}} / T_{\text{b final}}\right)} - \frac{T_{\text{ms}} - T_{\text{b final}}}{wL_0} \right] y_t \qquad (11.11)$$

Numerical Example Using Excel File *TLP-Risers.xls*

The Excel file *TLP-Risers.xls* gives the curves of riser excess setdown and riser stretch (given by eqq. [11.8] and [11.4], respectively) as a function of tension factor T_t/wL_0. Their intersection is found by iteration. The riser final profile is then plotted, and the position and value of the maximum riser sag is calculated using equations (11.10) and (11.11).

Seven data items are required for the calculations, as shown in table 11–1. The total apparent weight (wL_0) takes into account the apparent weight of the riser and all contents, including tubings, between the riser top end and the seabed. The riser is considered to be of constant apparent weight per unit length.

The specified value of tensioner stiffness in table 11–1 is very large, to simulate a locked-off riser without tensioners. The tendon length L_T is deduced from L_0 and L_P (see equation [11.5]).

Table 11–1. *Example data required for* TLP-Risers.xls

Top end offset, y_t (m)	50 and 75
Riser initial length, L_0 (m)	900
Platform column height, L_P (m)	62
Riser total apparent weight, w_{L0} (kN)	725
Initial tension factor, $T_{t\ initial}/w_{L0}$	1.3 and 1.9
Riser stiffness, $k_{riser} = \Sigma(EA/L)$ (kN/m)	480
Tensioner stiffness, k_{tens} (kN/m)	100,000

For compactness, table 11–1 summarizes data for four cases. It includes two values of top-end offset (y_t) and two values of the initial tension factor $T_{t\,initial}/wL_0$ for the vertical riser. When the file *TLP-Risers.xls* is used, only one value of each should be input.

Figure 11–3 shows the curves of riser excess setdown and riser stretch as a function of the top tension factor T_t/wL_0. The riser excess setdown depends on only the data in the first three lines of table 11–1. For the example, the right-hand side of equation (11.8) is equal to zero for T_t/wL_0 = 1.65. Hence, all the riser excess setdown curves pass through that point on the *y*-axis. That value of T_t/wL_0 depends on only the ratio L_0/L_T.

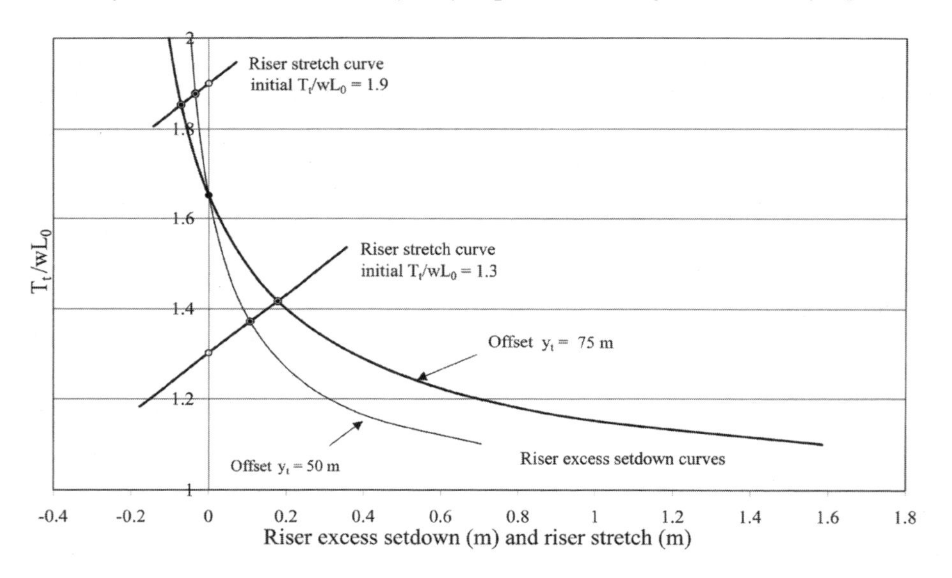

Fig. 11–3. *Example setdown and stretch curves for TLP risers*

The riser stretch curves depend on the data in last four lines of table 11–1. In figure 11–3, the intersection points between the riser excess setdown curves and the riser stretch curves give the final values of the tension factor $T_{t\,final}/wL_0$ and the riser stretch for the four specified cases.

For these examples, if the initial tension factor $T_{t\,initial}/wL_0$ is less than 1.65, the top tension will increase with platform offset, and the riser will stretch. For a top-end offset of 75 m and an initial tension factor of 1.3, it increases to a final factor $T_{t\,final}/wL_0$ = 1.42. If the initial tension factor is greater than 1.65, the top tension will decrease with offset, and the riser will shorten.

For the second tension factor, a large value (1.9) has been chosen as an illustration. For a top-end offset of 75 m, the tension factor is then seen to decrease to $T_{t\,final}/wL_0 = 1.85$. Figure 11–4 shows the resulting maximum sag, for a top-end offset of 75 m, for initial tension factors 1.3 and 1.9.

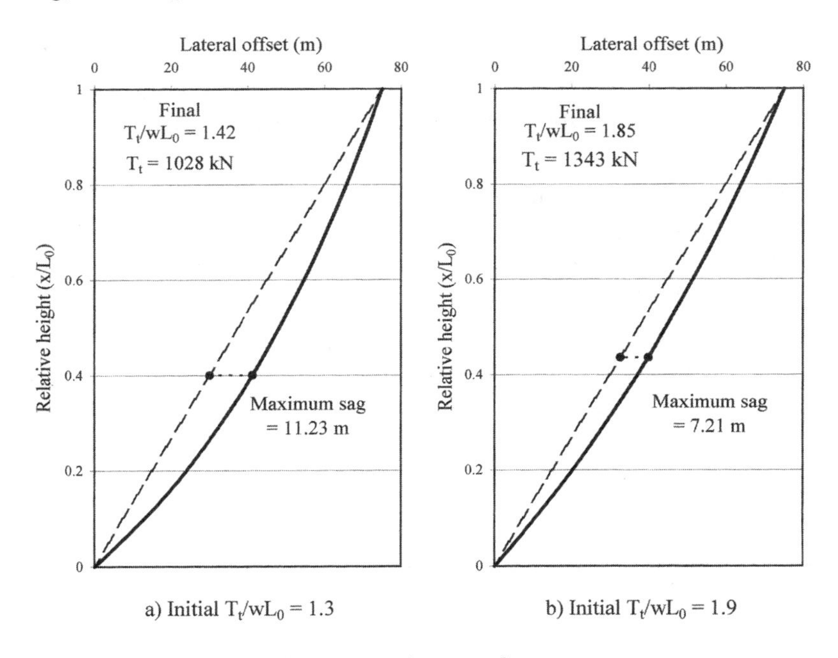

Fig. 11–4. *TLP riser sag for two initial tension factors*

The preceding calculations of riser offset-induced top tension, stretch, setdown, and sag can be carried out more accurately using a riser code capable of taking into account the third-order effects already mentioned, as well as the riser bending stiffness. Nevertheless, the same tendencies should be found.

Floating Platform Riser Tension and Sag Due to Offset: A Simplified Calculation

For a floating platform, the setdown depends principally on tide and platform weight and ballast. If the platform draught is little influenced by changes in riser tension, the platform setdown will no longer depend on top-end offset (y_t). Calling the top tension $T_{t\ nominal}$ for the platform at a *datum level,* the initial top tension $T_{t\ initial}$ for the vertical riser in the vertical position will then be reduced by the setdown z_{sd} with respect to that datum, according to the following relation:

$$T_{t\ initial} = T_{t\ nominal} - z_{sd} k_{r+t} \qquad (11.12)$$

where k_{r+t} is again the combined stiffness of the riser plus tensioners (see equation [5.44]).

It is important to verify that this reduction in top tension will not already cause instability or unacceptable bending at the lower end of the riser in the vertical position. It is generally sufficient to check that the riser bottom-end tension remains positive.

By use of the initial top tension $T_{t\ initial}$ given by equation (11.12), the total stretch is given, as before, by equation (11.4). However, the riser excess setdown has to be rewritten as equation (11.13), since the setdown z_{sd} is now independent of the platform offset y_t:

$$\text{riser excess setdown} = \frac{y_t^2 / 2L_0}{(T_t / wL_0)(T_b / wL_0)\left[\ln(T_t / T_b)\right]^2} - z_{sd} \qquad (11.13)$$

Numerical Example Using Excel File *Floater-Risers.xls*

The Excel file *Floater-Risers.xls* gives the curves of riser excess setdown and riser stretch, with respect to the platform *datum level,* as a function of the tension factor T_t/wL_0 according to equations (11.13) and (11.4), respectively. Their intersection is found by iteration. *Floater-Risers.xls* also plots the riser final profile and the position and value of the riser maximum sag calculated using equations (11.10) and (11.11).

Seven data items are required, as shown in table 11–2. Again, the total apparent weight (wL_0) must take into account the apparent weight of the riser and all contents, including tubings, between the riser top end and the seabed. The riser apparent weight per unit length is considered to be uniform between the platform deck and the seabed.

The specified value of tensioner stiffness in table 11–2 is very large to simulate a locked-off riser without tensioners. A very low value of riser axial stiffness k_{riser} is specified to simulate a riser made of new materials, such as high-performance composite.

Table 11–2. *Example data required for* Floater-Risers.xls

Top end offset, y_t (m)	55, 72, 85
Riser total nominal length, L_0 (m)	1200
Riser total apparent weight, w_{L0} (kN)	600
Nominal tension factor, $T_{t\ nominal}/w_{L0}$	1.35
Riser stiffness, $k_{riser} = \Sigma(EA/L)$ (kN/m)	115
Tensioner stiffness, k_{tens} (kN/m)	100,000
Floater setdown, z_{sd} (m)	1

Table 11–2 includes three values of top-end offset y_t. When using *Floater-Riser.xls,* only one value should be input. Figure 11–5 shows the riser excess setdown curves and riser stretch curves as a function of the top tension factor T_t/wL_0, according to equations (11.13) and (11.4), respectively.

The riser excess setdown curves depend on only the data given in the first two lines and the last line of table 11–2. Figure 11–5 also gives the initial tension factor $T_{t\ initial}/wL_0$ for the vertical riser with the platform setdown by z_{sd}, which causes the nominal tension factor (1.35) to be reduced to the initial tension factor (1.16). Since this is greater than 1, the bottom-end tension remains positive. The setdown is therefore acceptable.

Note that *Floater-Risers.xls* can also be used to make simplified calculations of risers tensioned from spars and other deep draught floaters. The data in the first four lines of table 11–2 then apply to the *platform keel level*; wL_0 is the riser apparent weight between the platform keel and the seabed, and the top tension $T_{t\ nominal}$ is the tension at the keel level, for the vertical riser with the platform at the datum level. By contrast, k_{riser} is the stiffness of the complete riser, including the part between the platform deck and keel. Tensioner stiffness k_{tens} and platform setdown z_{sd} are input as before.

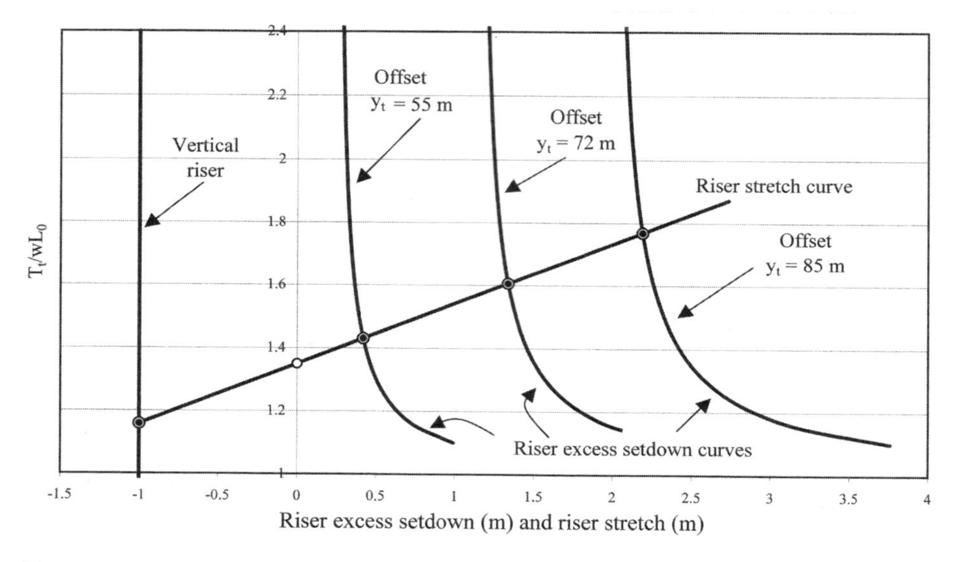

Fig. 11–5. *Example setdown and stretch curves for floating platform risers*

Influence of Internal Changes on Riser Tension and Profile

Internal changes of pressure, temperature, and fluid density will generally modify the tension and sag of a riser associated with a TLP or another floating platform. To calculate the consequences of the internal changes, it is simply necessary to introduce the modified data into table 11–1 or 11–2 and repeat the calculation. This involves calculating the changes in the apparent weight and in the top tension of the vertical riser. Note that the riser axial stiffness is unchanged by internal parameter changes.

The recalculation of the modified apparent weight is straightforward. There will be no change unless internal fluid densities change. The change in top tension (initial or nominal) of the vertical riser is more difficult to obtain. The calculation must take into account the fact that the initial platform level is unaltered. Hence, there will be no change in elongation of the vertical riser (i.e., $\Delta e = 0$).

The procedures for calculating the influence of internal changes on the relationship between riser top tension and axial elongation, for a vertical multi-tube riser, have been explained in detail in chapter 5. Since $\Delta e = 0$, equations (5.37)–(5.40) can be adapted to give the change in top-tension components (ΔT_t) of individual tubes, with or without expansion joint, in terms of the changes in apparent weight, internal and external pressures, and temperature (Δw_a, Δp_i and Δp_e, and Δt, respectively). When this is done, the definition of *components* of effective tension and apparent weight for multi-tube risers must be carefully respected (see chapter 3, bullet points following equations [3.10] and [3.13]).

For a uniform fixed tube, the change in top-tension component ΔT_t is given by (c.f. chapter 5, equation [5.19] and following):

$$\Delta T_t = \left(\Delta w_a \frac{L}{2} \right) - \left[(1 - 2\overline{\nu}_i) \Delta p_i A_i - (1 - 2\overline{\nu}_e) \Delta p_e A_e + (EA) \alpha \Delta t \right]_m \quad (11.14)$$

where L is the length of the tube and the subscript "m" implies midpoint values of Δp_i, Δp_e, and Δt. Equation (11.14) is written for anisotropic tubes, with internal and external equivalent Poisson's ratios ($\overline{\nu}_i$ and $\overline{\nu}_e$) since the equation is more general. Note that for isotropic tubes, $\overline{\nu}_i = \overline{\nu}_e = \nu$.

For a uniform tube with an expansion joint, the change in top-tension component ΔT_t is given by (c.f. chapter 5, equation [5.37]):

$$\Delta T_t = \Delta w_a d - \left(\Delta p_i A_J - \Delta p_e A_J \right)_{EJ} \quad (11.15)$$

where d is the depth of the expansion joint below the riser top end, A_J is the cross-sectional area of the expansion joint seal, and subscript "EJ" implies expansion joint–level values of Δp_i and Δp_e. Hence, the pressure- and temperature-induced changes in top tension of the vertical riser can be obtained. The modified values of $T_{t\,initial}/wL_0$ or $T_{t\,nominal}/wL_0$, for a TLP riser or a floating platform riser, can then be calculated, and the corresponding data table can be completed (table 11–1 or 11–2, respectively).

The influence on riser profile and sag will depend on the riser characteristics—in particular, on its elasticity and Poisson's ratio. Generally, the greater the elasticity is, the greater will be the change in riser sag. A possible exception is described in the next section. Changes in profile and sag must be carefully checked for acceptability, especially if they could lead to clashing with other risers.

Changes in apparent weight will always influence the riser profile, unless the riser is axially totally rigid. Thermal and internal pressure changes will generally also influence the riser profile, but it is possible to avoid them by using special materials and techniques, as explained in the next section.

Maximum sag, before and after parameter changes, will not occur at precisely the same point along the riser. However, the change in the value of the maximum sag should give an indication as to whether the design nominal spacing between adjacent risers is sufficient.

The stability of the vertical riser can be verified by checking that the bottom-end tension remains positive. Note that if the bottom-end tension is allowed to fall close to zero, the danger of catastrophic failure depends on how the riser is tensioned. If the riser top end is locked off, the top end will be *displacement controlled*; even if the tension at the lower end falls to zero, it will result only in increased bending, which may be acceptable. By contrast, if the riser is equipped with tensioners, the top end will be *load controlled*; if the bottom-end tension falls to zero, the riser top end will descend until the tensioners bottom out, resulting in potentially catastrophic uncontrolled buckling at the lower end.

Application to Composite Riser with Steel Tubings

This section examines a particular application consisting of a composite riser casing with dual completion. As explained previously, the influence of internal changes on riser profile and sag can be calculated once the modified top tension of the vertical riser has been determined. For the purpose of example, one tubing is fixed at extremities, while the other is equipped with an expansion joint. The calculation of the change in top tension owing to changes in temperature and internal pressures is explained using equations (11.16)–(11.18). Apparent weight is assumed to be constant, hence $\Delta w_a = 0$.

Composite risers have characteristics of low weight, high strength, and low axial stiffness, which theoretically allow them to be locked off to a platform without tensioners. It is also possible to design them for zero axial stretch when subject to changes of temperature and internal pressure, by designing for $\overline{\nu}_i = 0.5$ and zero coefficient of axial thermal expansion ($\alpha = 0$).[5]

For risers composed of uniform tubes, the changes in the top-tension components, owing to changes in temperature and pressure, can be derived from equations (11.14) and (11.15) (with $\Delta w_a = 0$). By assignment of the subscripts "casing" for the composite casing, "fixed" for the fixed tubing, and "free" for the tubing with an expansion joint, the changes in the top tension components of the three tubes are given as follows:

$$\Delta T_{\text{t casing}} = 0 \qquad (11.16)$$

$$\Delta T_{\text{t fixed}} = -\left[(1-2\nu)\Delta p_{\text{i}} A_{\text{i}} - (1-2\nu)\Delta p_{\text{a}} A_{\text{e}} + (EA)\alpha\Delta t\right]_{\text{m fixed}} \qquad (11.17)$$

$$\Delta T_{\text{t free}} = -\left(\Delta p_{\text{i}} A_{\text{J}} - \Delta p_{\text{a}} A_{\text{J}}\right)_{\text{EJ free}} \qquad (11.18)$$

where subscripts "m" and "EJ" imply midpoint values and values at the expansion joint, respectively (as for eqq. [11.14] and [11.15]).

The principal points can be summarized as follows:

- The riser top end is fixed to the platform. Hence, there will be no change in axial elongation of the vertical riser ($\Delta e = 0$).
- For the composite casing, $\overline{\nu_{\text{i}}} = 0.5$, $\alpha = 0$, and $\Delta p_{\text{e}} = 0$ (i.e., constant external pressure). When input into equation (11.14), these values give zero change in the top-tension component ($\Delta T_{\text{t casing}}$), as given by equation (11.16).
- For the tubings, Δp_{a} is the change in annular pressure.
- For the fixed tubing, changes in temperature or pressure (internal or annular) will change the top-tension component ($\Delta T_{\text{t fixed}}$), as given by equation (11.17).
- For the tubing with an expansion joint, changes in pressures (internal and annular) will change the top-tension component ($\Delta T_{\text{t free}}$), as given in equation (11.18).

Both tubings lead to pressure- and temperature-induced changes in top tension and, hence, to changes in profile and sag. The fixed tubing will greatly increase the riser total axial stiffness. This is unacceptable since it destroys the benefit of the axial flexibility of the composite casing and causes the riser tension to be taken principally by the tubing.

The solution with an expansion joint may lead to large pressure-induced changes in the top tension (see equation [11.18]), with large changes in riser profile and sag. This can be avoided by using a special, *balanced* expansion joint as explained in the following subsection.

Tubing with balanced expansion joint

This subsection describes an expansion joint for which the effective tension component remains constant whatever the internal and external pressures. Since such a joint is particularly pertinent to tubings for risers with stiff tensioners, it is described here rather than in chapter 3.

The object is to design a special expansion joint for which the effective tension is always zero, whatever the internal and external pressures. Since the top-tension component of a tubing is equal to the effective tension in the joint, plus the apparent weight of the segment of tubing between the joint and the top end, the top-tension component will remain constant providing that there is no change in apparent weight.

Figure 11–6 shows a balanced expansion joint, which consists of a standard expansion joint surrounded by a double pressure chamber. The joint and the chamber include three sliding annular seals. In figure 11–6, the upper chamber communicates with the interior of the tube via an internal port, while the lower chamber communicates with the annulus via the external port.

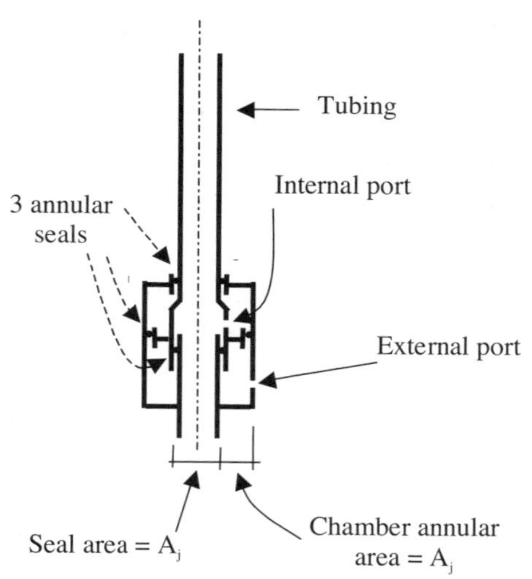

Fig. 11–6. *A balanced expansion joint*

The annular cross-sectional area of the chamber around the joint is designed to be equal to the seal area A_j. In figure 11–6, the internal pressure p_i in the upper part of the chamber causes an axial tension in the tubing wall equal to p_iA_j. Likewise, the external (annular) pressure p_a in the lower part of the chamber causes an axial compression in the pipe wall equal to p_aA_j. Hence, the chamber contributes a true wall tension at the joint (immediately above it) equal to the sum of the two forces: $T_{tw} = p_iA_j - p_aA_j$. However, the effective tension at an expansion joint *without* chamber is given by $T_e = -p_iA_j + p_aA_j$ (see equation [5.36]). The effective tension for the balanced expansion joint is given by equation (11.19) (cf. equation [2.19]), which is always zero, whatever the pressures:

$$T_e = \left(p_iA_j - p_aA_j\right) - p_iA_j + p_aA_j = 0 \qquad (11.19)$$

Thus, for a tubing with a balanced expansion joint, the top-tension component will be constant, and changes in pressure and temperature will have no influence.

Influence of tubing pressure on riser profile

The preceding arguments have shown that tubing pressure changes will influence riser top tension and, hence, riser profile and sag, unless the tubing is equipped with a balanced expansion joint. Some engineers have been surprised that a pressure change in a tubing can have any influence at all on riser profile, particularly if the tubing is equipped with a (standard) expansion joint deep in the well.

The problem is very similar to that discussed in chapter 3 (see fig. 3–3). The axial force in the internal fluid column in the tubing creates a lateral thrust that acts on the curved tubing. This force is transmitted to the riser casing via the centralizers and hence influences the riser profile.

Summary

In this chapter, the relationship between platform offset and setdown, as well as riser stretch, has been explained both for TLP risers and floating platform risers. Expressions for the influence of platform offset and setdown on riser tension and sag have been derived, based on the simple expressions of chapters 7. Those expressions have been applied in the Excel files *TLP-Risers.xls* and *Floater-Risers.xls,* which have been used to obtain numerical results discussed within this chapter.

The influences that internal changes (to pressures, temperatures, and internal fluid densities) exert on riser tension have been discussed. A method for calculating the overall influence of those changes on riser profile and sag has been explained. The method requires the modified apparent weight and the top tension of the vertical riser to be determined first, taking into account the stiffness of the connection between the riser and the platform. The method can be applied to multi-tube risers.

The particular case of a composite riser casing with steel tubings has been examined in detail. The special characteristics of composites allow such a casing to be locked off to the platform without tensioners, without being subject to pressure- and temperature-induced changes in riser tension or sag. However, changes in tubing pressures (and temperature in the case of a fixed tubing) do normally influence the riser profile. It has been shown that this can be avoided by equipping the tubing with a carefully designed balanced expansion joint. The way in which tubing pressure can influence riser profile has been explained.

References

[1]Petersen, W. H., R. W. Patterson, J. D. Smith, E. B. Denison, D. W. Allen, A. G. Ekvall, E. H. Phifer, and Y. S. Li. 1994. Auger TLP well systems. Paper OTC 7617, presented at the Offshore Technology Conference, Houston;
Alfstad, O., P. Rooney, O. Dankertsen, and B. Manning. 1991. Snorre riser and well systems. Paper OTC 6624, presented at the Offshore Technology Conference, Houston;
Denison, E. B., C. T. Howell, G. T. Ju, R. Gonzalez, G. A. Myers, and G. T. Ashcombe. 1997. Mars TLP drilling and production riser systems. Paper OTC 8514, presented at the Offshore Technology Conference, Houston;
Carminati, J. R., L. F. Eaton, K. A. Folse, and R. E. Sokoll. 1999. URSA TLP well systems. Paper OTC 10758, presented at the Offshore Technology Conference, Houston.

[2]Berner, P., V. Baugus, K. Gendron, and R. Young. 1997. Neptune Project: Production riser system design and installation. Paper OTC 8386,

presented at the Offshore Technology Conference, Houston;
Perryman, S., J. Gebara, F. Botros, and C. A. Yu. 2005. Holstein truss SPAR and top tensioned riser system design challenges and innovations. Paper OTC 17292, presented at the Offshore Technology Conference, Houston.

[3]Huang, E., S. Bhat, Y. Luo, and J. Zou. 2000. Evaluation of dry tree platform concepts. Paper OTC 11899, presented at the Offshore Technology Conference, Houston;
Wanvik, L., and J. M. Johnsen. 2002. Deep water moored semisubmersible with dry wellheads and top tensioned well risers. Paper OTC 12986, presented at the Offshore Technology Conference, Houston;
Poldervaart, L., and J. Pollack. 2002. A dry tree FDPSO unit for Brazilian waters. Paper OTC 14256, presented at the Offshore Technology Conference, Houston;
Gupta, H., L. Finn, and J. Halkyard. 2002. SPAR riser alternatives for 10,000 ft water depth. Paper OTC 14298, presented at the Offshore Technology Conference, Houston;
Bhat, S. U., W. L. Griener, and D. Barton. 2003. Deepstar 10,000 ft water depth DDCV study. Paper OTC 15102, presented at the Offshore Technology Conference, Houston.

[4]Sparks, C. P., P. Odru, H. Bono, and Métivaud. 1988. Mechanical testing of high-performance composite tubes for TLP production risers. Paper OTC 5797, presented at the Offshore Technology Conference, Houston;
Baldwin, D. B., and R. J. Long. 1999. Mechanical performance of composite production risers. Paper OTC 11008, presented at the Offshore Technology Conference, Houston.

[5]Sparks, C. P., and J. Schmitt. 1990. Optimized composite tubes for riser applications. Paper presented at the 9th Offshore Mechanics and Arctic Engineering Conference, Houston.

12

Steel Catenary Risers

Steel catenary risers (SCRs) have been associated with floating platforms since the mid-1990s and were first used as export risers for the Auger TLP.[1] Since then, they have been used progressively for more severe applications. They have been used as export risers for semisubmersibles and as production risers for FPSOs.[2] Such risers might be better called *stiff* catenary risers, since there are serious projects for building them from materials other than steel in the future.[3] SCRs are the subject of much ongoing research, particularly with respect to fatigue, interaction with the seabed, and vortex-induced vibrations (VIV).[4]

Note that a catenary is in fact a precise mathematical curve, but the term is used loosely in the offshore industry to refer to any riser that approximates to that mathematical curve. Throughout this chapter, for clarity, the expression *stiff catenary* is used to refer to a catenary riser with bending stiffness EI, and *cable catenary* is used to refer to one without bending stiffness ($EI = 0$).

Note again that all references to *tension* imply effective tension.

Basic Differential Equation

SCRs are horizontal at the lower end and generally within about 20° of the vertical at the top end. Hence, their total profile and curvature cannot be even remotely analyzed using small-angle deflection theory. They can, of course, be analyzed by defining local axes for a number of sections (10 or more) for which the angles do not evolve by more than 10°, which is the limit normally accepted for small-angle deflection theory. It is, however, simpler to abandon (x, y) coordinates and use (θ, s) coordinates instead, as defined in figure 12–1, where θ is the angle with the horizontal and s is measured from the touchdown point (TDP).

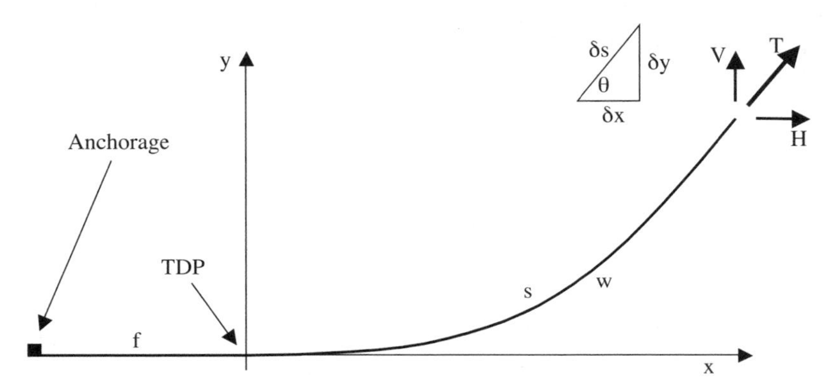

Fig. 12–1. *Catenary axes and symbols*

The basic differential equation governing curvature and deflections of tensioned beams subject to large deflections is derived in appendix A (see equations [A.9]–[A.11]). For a beam with uniform bending stiffness, the basic equation is as follows:

$$EI\frac{d^3\theta}{ds^3} - T\frac{d\theta}{ds} + w\cos\theta + f(s) = 0 \qquad (12.1)$$

where $w \cos\theta$ and $f(s)$ are the respective components of the self weight and the in-plane current load, acting perpendicular to the catenary axis; T is the tension. Since the curvature is given by $1/R = d\theta/ds$, equation (12.1) can be rewritten as

$$EI\frac{d^2}{ds^2}\left(\frac{1}{R}\right) - T\left(\frac{1}{R}\right) + w\cos\theta + f(s) = 0 \qquad (12.2)$$

Cable Catenary Equations

The cable catenary has zero bending stiffness and zero current load. Equation (12.1) then has a standard analytical solution, which can be found in textbooks. For completeness, the equations are rederived in appendix H (see equations [H.1]–[H.10]) and are summarized here by equations (12.3)–(12.5):

$$\frac{T}{H} = \frac{wy}{H} + 1 = \cosh\left(\frac{wx}{H}\right) \tag{12.3}$$

$$\frac{ws}{H} = \frac{dy}{dx} = \sinh\left(\frac{wx}{H}\right) \tag{12.4}$$

$$\frac{1}{R} = \frac{wH}{T^2} \tag{12.5}$$

Curvature ($1/R$) is maximum at the TDP ($x = 0$, $s = 0$), where it is equal to w/H. This should be no surprise since the part of the catenary adjacent to the TDP can be treated as a horizontal cable subject to vertical load w and axial tension H.

Equations (12.3)–(12.5) have the merit of great simplicity. However, they neglect the effects of axial stretch and bending stiffness. The equations can be used to estimate the shift of the TDP resulting from the riser top-end movements, which may be very large.

TDP Shift Due to Top-End Movement Using Excel File *TDP-Shift.xls*

Floating platforms can be subject to large static lateral displacements as a result of wind, current, and other loads. When these displacements are in the plane of an SCR, they may result in large changes in the position of the TDP, as illustrated by figure 12–2.

If the riser is analyzed as a simple, inextensible cable, calculation of the TDP shift is straightforward by use of equations (12.3) and (12.4). In figure 12–2, the suspended length s_t and its horizontal projection x_t are measured from the TDP. The total length S_t of the catenary plus flow line is constant, given by $S_t = s_t + f$, where f is the length of the flow line. The total horizontal projection X_t is given by $X_t = x_t + f$, and the shift in position of the TDP is equal to the change in flow line length f.

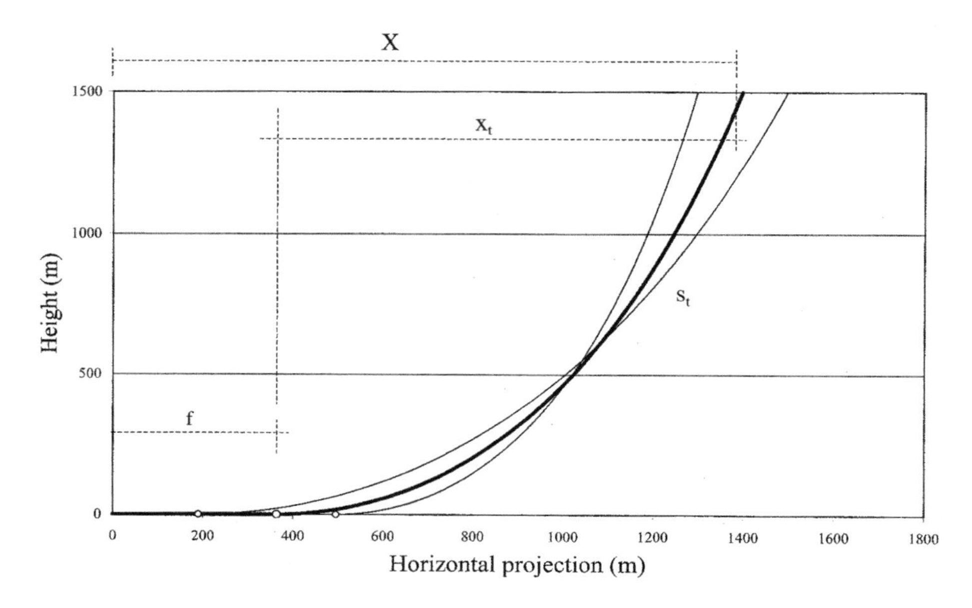

Fig. 12–2. *TDP shift as a result of top-end lateral displacement*

Any change in the horizontal projection X_t will cause a change in the top tension T_t. An iterative calculation will then be required to find the modified value. The Excel file *TDP-Shift.xls* can be used to explore the influence of the amplitude of the top-end movement on TDP shift. For ease of printing, an alternative version of the file, named *TDP-Shift (print).xls* is also provided.

Table 12–1. *Example data for TDP shift calculation*

Depth (m)	1,000
Total length, S_t (m)	1,550
Apparent weight per unit length, w (kN/m)	1.6
Top tension, T_t (kN)	2,150
Amplitude of top end movement (m)	75

Table 12–1 gives example data required for a typical calculation. Table 12–2 gives the corresponding results.

Table 12–2. *Example results obtained with* TDP-Shift.xls

Results	Position		
	Near	Mean	Far
Top end movement (m)	-75	-	+75
Top tension, T_t (kN)	1,951	2,150	2,447
Horizontal force, H (kN)	351	550	847
Catenary: suspended length, s_t (m)	1,200	1,299	1,435
horizontal projection, x_t (m)	527	701	912
Seafloor flowline length, f (m)	350	251	115
Total horizontal projection, $X_t = x_t + f$ (m)	877	952	1,027
TDP shift (m)	100	-	-136
Top end angle with vertical	10.4°	14.8°	20.3°
Max. curvature (at TDP) (m^{-1})	0.0046	0.0029	0.0019

As shown in table 12–2, *TDP-Shift.xls* also gives the changes in top tension T_t, horizontal force H, and top-end angle with the vertical, for the different top-end positions. The file also gives the natural periods of *transverse modal vibrations,* which are explained in chapter 15.

Catenary and Flow-Line Stretch

It may be important to know the riser stretch when it is initially deployed and to know the stretch induced by parameter changes, such as changes in temperature, pressure, and internal fluid density. The stretch may be small for SCRs, made from steel, but could be more significant for more elastic materials.

Expressions for catenary axial stretch Δs and its projections Δx and Δy are derived and explained in detail in appendix H. A simple way of estimating the significance of the stretch is given in the next subsection.

The axial stretch can be found by integrating the axial strain $\Delta\varepsilon_a$, given by equation (12.6) (see equations [H.14] and [5.7]), along the length of the catenary and seafloor flow line:

$$\Delta\varepsilon_a = \frac{1}{EA}\left[\Delta T + \left(1-2\overline{\nu}_i\right)\Delta p_i A_i - \left(1-2\overline{\nu}_e\right)\Delta p_e A_e\right] + \alpha\Delta t \quad (12.6)$$

where ΔT, Δp_i and Δp_e, and Δt are the respective changes in effective tension, internal and external pressure, and temperature at a particular point, between two sets of conditions; α is the coefficient of thermal expansion; A_i and A_e are the internal and external cross-sectional areas; and EA is the axial stiffness. Equation (12.6) is given for anisotropic pipes in terms of the internal and external equivalent Poisson's ratios $\overline{\nu_i}$ and $\overline{\nu_e}$, since it is more general. For isotropic pipes, $\overline{\nu_i} = \overline{\nu_e} = \nu$.

It is shown in appendix H that a parameter B can be defined by

$$B = \frac{\left(1-2\overline{\nu_i}\right)\rho_i g A_i - \left(1-2\overline{\nu_e}\right)\rho_e g A_e}{w} \qquad (12.7)$$

where ρ_i and ρ_e are internal and external fluid densities, g is the gravitational constant, and w is the apparent weight per unit length. The strain ε_a at any point owing to deployment can then be expressed in terms of the height y above the seabed:

$$\varepsilon_a = \frac{1}{EA}\left[T + \left(1-2\overline{\nu_i}\right)p_{ti}A_i + Bw(d-y)\right] + \alpha t \qquad (12.8)$$

where T is the effective tension at the point, p_{ti} is the top-end final internal pressure, t is the change in temperature owing to deployment, and d is the total depth. Total stretch is found by integrating equation (12.8) along the length of the riser. This is done in appendix H.

Following deployment, the total catenary stretch Δs, its projections Δx and Δy, and the flow line stretch Δf (ignoring soil friction) are given in appendix H by equations (H.30), (H.23), (H.25), and (H.33), respectively, assuming that the pipe is initially horizontal, at a datum temperature, and subject to zero tension and pressures. The influence of parameter changes can be found by evaluating Δs, Δx, Δy, and Δf, before and after the parameter changes, and then taking the difference.

Estimate of total stretch ($\Delta s + \Delta f$)

Since the equations of appendix H may appear complicated, a quick way of determining whether the axial stretch is significant may be helpful. The total riser stretch can be obtained from the mean axial strain, which can be estimated using a mean value of the effective tension T in equation (12.8). All other parameters in equation (12.8) are constant, including B (given by equation [12.7]).

The total stretch $\Delta s + \Delta f$ will then be equal to that mean strain times the total unstretched length S_u. Using a mean value of effective tension $T = (T_t + H)/2$ in equation (12.8) gives a quick approximation of the *mean strain* and, hence, the *mean stretch*. A more accurate estimate can be obtained by using $T = (T_t + H)/2$ as the mean effective tension for the suspended length and H for the seabed flow line.

Global Influence of Bending Stiffness

In chapter 6, it was shown that the curvature of a tensioned horizontal beam is close to that of a tensioned cable, except for zones close to the extremities. In chapter 8, the same was found to be true for near-vertical risers.

It has not been found possible to write a simple Excel file that allows the curvature and profile of catenaries, with and without bending stiffness, to be immediately compared, as it was for beams and risers. However, results of static simulations made with standard computer programs can easily be used to show that bending stiffness also has little influence on catenary curvature except for zones close to the TDP and top end.

To illustrate points raised in this chapter, several simulations have been made of catenaries with different values of bending stiffness. The basic data used are summarized in table 12–3.

In table 12–3, the total length includes a horizontal length of flow line on the seabed. The total horizontal projection, of the suspended length plus the seabed flow line, was maintained constant for all the simulations. The corresponding top tension was then found for values of EI, which was made to vary between zero and 4,000,000 kN-m^2. For all cases, the catenaries were inextensible, and the current load was zero.

Table 12–3. *Basic data for example catenary simulation*

Water depth (m)		1,500
Total length, S_t (m)		2,030
Total horizontal projection, X_t (m)		1,125.23
Top tension, T_t (kN)		2,000 to 1,990
Apparent weight per unit length, w (kN/m)		1
Current load, f(s) (kN/m)		0
Axial elasticity, 1/EA (kN^{-1})		0
Bending stiffness, EI (kN-m^2)		0 to 4 × 10^6
Element length, e (m):	- upper 1,900m	5
	- remainder	2
Soil compressibility		Incompressible

The catenary curvature is plotted in figure 12–3 as a function of arc length (measured from the top end), for five values of *EI*: zero, 10^5, 5 × 10^5, 2 × 10^6, and 4 × 10^6 kN-m^2. Cable curvature is shown in boldface.

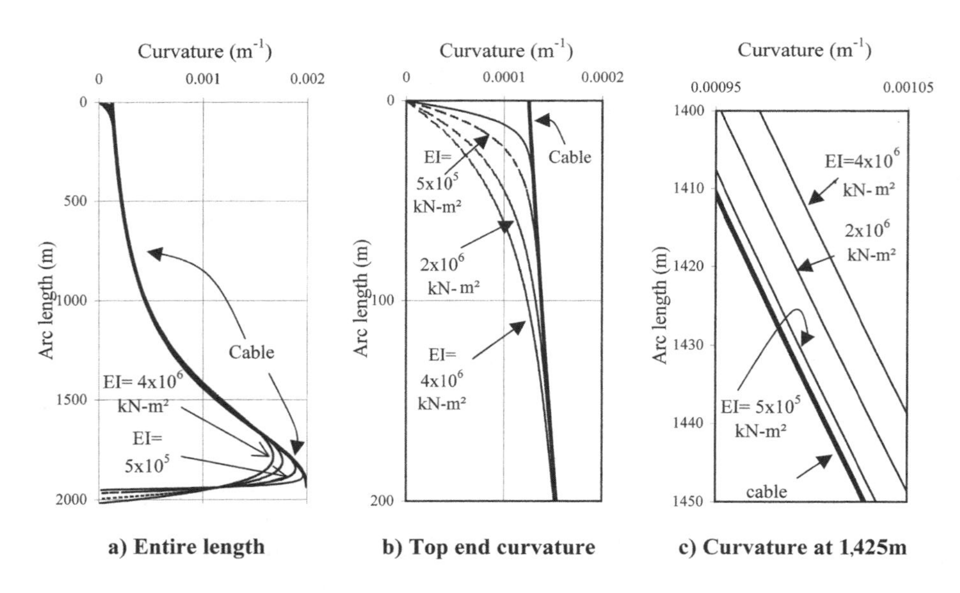

Fig. 12–3. *Curvature as a function of arc length, for different values of EI*

Figure 12–3*a* shows the curvature for the five catenaries over the entire length. As expected, curvature is very close to cable curvature everywhere except for zones close to the extremities. Some confusion at the lower end results because the TDPs are at different arc lengths from the top end.

Figure 12–3*b* shows the top-end curvature in detail. Curvature there is small since tension is high and the angles are close to the vertical. Hence, the lateral component of self weight is small. For each catenary, curvature converges toward cable curvature as distance from the extremity increases, as was seen for near-vertical risers in chapter 8.

Figure 12–3*c* shows the curvature over a short length in the central zone, for different values of *EI*. The curvatures and their differences are small, but in all cases, curvature is *greater* for the stiff catenaries than for the cable catenary; the greater the stiffness is, the greater will be the curvature! This tendency dominates over approximately 70% of the total length of the catenary. Although this appears to contradict engineering intuition, it results from a small redistribution of load induced by the shear capacity of the stiff catenaries. This was already observed for horizontal tensioned beams in chapter 6 (see following equation [6.13]).

Further evidence of the small influence of bending stiffness on the catenary profile is provided by a comparison between angles at points along the different catenaries. Table 12–4 compares the angles—with the vertical—at different arc lengths from the top end for catenaries corresponding to the data of table 12–3, for seven values of *EI*.

Table 12–4. *Influence of bending stiffness on angles (with the vertical) at different arc lengths from the top end*

Arc length from top end	**Bending stiffness EI ($kN\text{-}m^2$)**						
	Cable	100,000	300,000	500,000	1,000,000	2,000,000	4,000,000
0	14.48°	14.53°	14.55°	14.56°	14.58°	14.59°	14.59°
300 m	16.99°	16.98°	16.97°	16.96°	16.93°	16.87°	16.77°
750 m	22.85°	22.85°	22.83°	22.82°	22.79°	22.74°	22.65°
1,500 m	48.88°	48.90°	48.93°	48.97°	49.06°	49.24°	49.59°
1,850 m	80.19°	80.22°	80.25°	80.26°	80.22°	80.08°	79.80°

Table 12–4 shows that the angles are very close to those of the cable catenary even for the very stiffest catenary (EI = 4,000,000 kN-m^2), equivalent to a steel pipe with 54 in. outer diameter and 3/4 in. wall thickness. It is particularly significant that there should be such close agreement down to the point where the angles are close to 80°, since the segment between that point and the TDP can be analyzed using small-angle deflection theory.

Details of Numerical Analyses

The relative influence of bending stiffness on catenary curvature and profile can also be appreciated by comparing the components of equation (12.1), obtained by numerical analyses. A numerical expression for equation (12.1) is derived in appendix H (see equations [H.34]–[H.42]). At any node n, as shown on figure 12–4, the terms of equation (12.1) can be expressed in terms of the element length e between nodes and the angles at the five nodes $n - 2$ to $n + 2$.

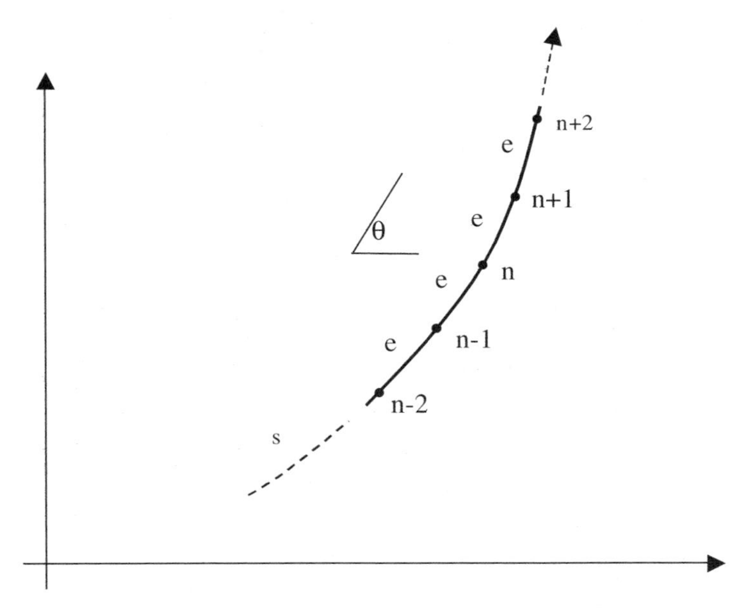

Fig. 12–4. *Equally spaced adjacent nodes on a stiff catenary*

For the case with zero current, equation (12.1) can be expressed numerically by equation (12.9), where T_n is the tension at node n:

$$EI\left(\frac{\theta_{n+2}-2\theta_{n+1}+2\theta_{n-1}-\theta_{n-2}}{2e^3}\right)-T_n\left(\frac{\theta_{n+1}-\theta_{n-1}}{2e}\right)+w\cos\theta_n=0 \qquad (12.9)$$

In equation (12.9), the influence of the bending stiffness EI is contained in the first term. The second term is equal to T_n/R_s, where $1/R_s$ is the curvature of the stiff catenary. The third term is equal to T_n/R_c, where $1/R_c$ is the curvature of the cable catenary. Hence, equation (12.9) can be rewritten as

$$EI\left(\frac{\theta_{n+2}-2\theta_{n+1}+2\theta_{n-1}-\theta_{n-2}}{2e^3}\right)-T_n\frac{1}{R_s}+T_n\frac{1}{R_c}=0 \qquad (12.10)$$

Evaluating the three terms of equation (12.9) individually allows the curvature of the stiff catenary to be compared with that of the cable catenary. It also reveals where the first and third terms have the same sign—and, hence, where the bending stiffness causes an *increase* in curvature. The sum (Σ) of the three terms, which should be zero, gives the precision of the numerical calculation.

Table 12–5. *Components of equation (12.1) for seven catenaries, at 1,425 m arc length from top end*

Bending stiffness EI (kN-m²)	**Equivalent (steel) OD and wall thickness**	**EI.d³θ/ds³**	**- T.dθ/ds**	**w.cos θ**	**Σ**
Zero	Cable	0	0.6990	-0.6990	0.0000000
100,000	18” x 9/16”	-0.0004	0.6998	-0.6994	0.0000035
300,000	22” x 15/16”	-0.0012	0.7009	-0.6997	0.0000034
500,000	26” x 15/16”	-0.0020	0.7020	-0.7000	0.0000035
1,000,000	32” x 1”	-0.0039	0.7047	-0.7007	0.0000035
2,000,000	42” x 7/8”	-0.0075	0.7098	-0.7023	0.0000036
4,000,000	54” x 3/4”	-0.0125	0.7181	-0.7056	0.0000034

For the seven simulations, table 12–5 gives the components of equation (12.1), calculated according to equation (12.9), at 1,425 m arc length from the top end, which is the region where the differences between the curvatures of the stiff and cable catenaries are greatest, apart from the extremities. As can be seen, the influence of bending stiffness is very small, even for the stiffest catenary. It can also be seen that equation (12.9) is solved numerically with great precision, since Σ is close to zero for all cases.

The Excel file *Catenary-500.xls* gives the three components of equation (12.9) *for all nodes,* for the case with EI = 500,000 kN-m², which allows the influence of the bending stiffness to be appreciated in detail over the entire length of the catenary. The value of T/R_c, calculated analytically according to equation (12.5) (by rearrangement as $T/R_c = wH/T$), and the ratio of curvatures of the stiff and cable catenaries (R_c/R_s) are also tabulated. Between arc lengths 125 and 1,635 m, the term $EI(d^3\theta/ds^3)$ is *negative,* and bending stiffness causes a slight *increase* in curvature.

Influence of Bending Stiffness on TDP position

Bending stiffness for the seven catenaries of table 12–4 was seen to have negligible influence on slope down to points within 10° of the horizontal. Note also from figure 12–3*a* that the curvature at that point (arc length 1,850 m) is still very close to cable curvature for all but the two stiffest catenaries.

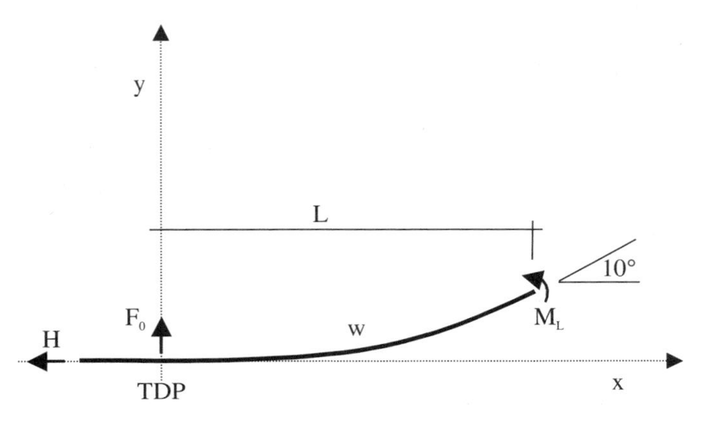

Fig. 12–5. *Catenary segment between TDP and 10° point*

For all but the stiffest catenaries, it should be possible to analyze the segment between the TDP and the 10° point, shown in figure 12–5, by using small-angle deflection theory and assuming that the catenary curvature is equal to cable curvature at the 10° point. The tensioned-beam equations of chapter 6 can then be applied directly.

The subscripts "0" and "L" are used to mean the TDP and the 10° point, as shown in figure 12–5. The load on the segment is uniform and equal to the apparent weight w, and the tension is constant and equal to H. The moment M_0 at the TDP of the stiff catenary is zero. At the 10° point, curvature is virtually equal to cable curvature; the moment M_L there can be deduced, since $M_L/EI = 1/R_{cable} = w/H$. Hence, from equations (6.8) and (6.9), $G_0 = -w/H$, and $G_L = 0$.

Substituting G_0 and G_L into equation (6.11) leads to

$$\left(\frac{1}{R}\right)_x = -\frac{w}{H}e^{-kx} + \frac{w}{H} \qquad (12.11)$$

where $k = \sqrt{H/EI}$ (see equation [6.4]). Since the curvature $1/R = d^2y/dx^2$,

$$\frac{d^2y}{dx^2} = \frac{w}{H}\left(1 - e^{-kx}\right) \qquad (12.12)$$

Integration of equation (12.12), with constant of integration chosen to give $dy/dx = 0$ at the TDP, leads to

$$\frac{dy}{dx} = \frac{w}{H}\left(x - \frac{1 - e^{-kx}}{k}\right) \qquad (12.13)$$

As mentioned in chapter 6, e^{-kx} becomes rapidly negligible for $kx > 3$ (see following equation [6.11]). Hence, equation (12.13) tends toward

$$\frac{dy}{dx} = \frac{w}{H}\left(x_s - \frac{1}{k}\right) \qquad (12.14)$$

where x_s is the distance measured from the TDP of the stiff catenary.

For a cable catenary, curvature between the TDP and the 10° point will be constant, equal to w/H. The slope is then given by

$$\frac{dy}{dx} = \frac{w}{H} x_c \qquad (12.15)$$

where x_c is the distance measured from the TDP of the cable catenary.

When points are chosen beyond the zone of influence of the extremity ($kx_s > 3$), where the slope is the same for both catenaries, equations (12.14) and (12.15) yield

$$x_s - \frac{1}{k} = x_c \qquad (12.16)$$

The distance $x_s - x_c$ is then the difference in position of the TDPs of the stiff catenary and the cable catenary. Since, from equation (6.4), $k = \sqrt{H/EI}$, the difference in position of the TDPs is given as follows:

$$x_s - x_c = \sqrt{\frac{EI}{H}} \qquad (12.17)$$

Table 12–6 includes the flow-line lengths of the cable catenary and six stiff catenaries, calculated numerically. The difference in position of the TDP, for each case, is equal to the difference in flow-line lengths of the stiff catenary as compared to the cable catenary. That difference is tabulated in the "Numerical result" column of table 12–6, where it is compared with the analytical value ($\sqrt{EI/H}$) given by equation (12.17). Agreement is seen to be good for bending stiffnesses up to EI = 1,000,000 kN-m², corresponding to a steel pipe with 32 in. outer diameter and 1 in. wall thickness.

For the two largest values of EI (2,000,000 and 4,000,000 kN-m²) in table 12–6, the errors in the analytical calculations are 7% and 11% respectively. This reduction in precision is not surprising since one of the assumptions in the preceding analysis was that the 10° point should be beyond the zone of influence of the extremity (i.e., $kx_s > 3$), which is not the case for the two stiffest catenaries.

The curvature ratio (R_c/R_s) between the stiff and cable catenary curvatures at the 10° point, is also given in table 12–6. As expected, the precision ratio is reduced as bending stiffness increases. The value of kx_s

at the 10° point can therefore be used as a good criterion for judging the precision of the TDP calculation.

Table 12–6. *Influence of bending stiffness on TDP position*

Bending stiffness EI (kN-m²)	Forces (kN)		Flowline length (m)	Difference in TDP (m)		kx_s at 10° point	Ratio R_c/R_s at 10° point
	T_t	H	f	Numerical result	Analytical √(EI/H)		
Cable	2,000	500	93.5	-	-	-	1
100,000	2,000.3	499.9	78	15.5	14.1	7.2	0.997
300,000	1,999.7	499.3	70	23.5	24.5	4.6	0.982
500,000	1,999.1	498.7	62	31.5	31.7	3.8	0.964
1,000,000	1,997.8	497.4	50	43.5	44.8	3.0	0.924
2,000,000	1,995.2	494.8	34	59.5	63.6	2.4	0.867
4,000,000	1,990.4	490.0	12	81.5	90.4	2.0	0.797

Top Tension, TDP Shear Force, and Soil Reaction

The value of the top tension T_t is also included in table 12–6 for each of the simulations. As given in table 12–3, the total length and the total horizontal projection were 2,030 m and 1,125.23 m, respectively, for all cases. Since the suspended length increases with the bending stiffness (see table 12–6), it may seem surprising that the top tension remains virtually unchanged, whatever the stiffness. The explanation lies in the shear force at the TDP.

For the coordinate system of figure 12–5, the shear force F in the near-horizontal segment close to the TDP is given by

$$F = \frac{dM}{dx} = \frac{d}{dx}\left[EI\left(\frac{d^2y}{dx^2}\right)\right] = EI\left(\frac{d^3y}{dx^3}\right) \qquad (12.18)$$

Differentiating equation (12.12) gives $d^3y/dx^3 = (w/H)ke^{-kx}$. Hence, the shear force close to the TDP is given by

$$F = EI\left(\frac{w}{H}\right)ke^{-kx} \qquad (12.19)$$

Since $k = \sqrt{H/EI}$ (cf. equation [6.4]), the TDP shear force, which is transmitted to the soil, is given by

$$F_{\mathrm{TDP}} = w\sqrt{EI/H} \qquad (12.20)$$

However, $\sqrt{EI/H}$ is the excess suspended length of the stiff catenary as compared to the cable catenary (see equation [12.17]). Thus, the shear force at the TDP is equal to the weight of that excess length. Therefore, the vertical component of the top tension V_{t} remains unchanged.

Because the horizontal force H is also unchanged (see table 12–6) and $T_{\mathrm{t}} = \sqrt{V_{\mathrm{t}}^2 + H^2}$, the top tension T_{t} is virtually unchanged. In table 12–6, this is still seen to be true, even for the two very stiff catenaries; this is surprising, given that the zone of influence of the lower extremity (i.e., the TDP) extends well beyond the 10° point.

Summary

In this chapter, the basic large-angle deflection equation and the solution for the case of a cable catenary with zero bending stiffness have been recalled. In the Excel file *TDP-Shift.xls,* the cable catenary equations have been applied to the calculation of movement of the TDP resulting from the catenary top-end movement. Expressions for the axial stretch owing to deployment of an SCR and to internal parameter changes have been derived and discussed.

The influence of bending stiffness has been examined, using the results of numerical simulations, for a wide range of bending stiffnesses. It has been shown that the influence on curvature is principally limited to the extremities, which agrees with the findings of chapters 6 and 8 for horizontal beams and near-vertical risers.

The influence of bending stiffness on the TDP position, top tension, TDP shear force, and soil reaction have been examined. Simple expressions have been formulated for the difference in position of the TDP of a stiff catenary as compared to a cable catenary. Other expressions have been derived for the shear force and, hence, the soil reaction at the TDP of a stiff catenary. Top tension of a stiff catenary has been shown to be virtually independent of bending stiffness and equal to the top tension of a cable catenary, even though the suspended lengths may be very different.

References

[1]Phifer, E. H., F. Koop, R. C. Swanson, D. W. Allen, and C. G. Langer. 1994. Design and installation of Auger steel catenary risers. Paper OTC 7620, presented at the Offshore Technology Conference, Houston.

[2]Serta, O. B., M. M. Mourelle, F.W. Grealish., S.J. Harbert, L.F.A. Souza. 1996. Steel catenary riser for Marlim Field FPS P-XVIII. Paper OTC 8069, presented at the Offshore Technology Conference, Houston.

[3]Nygard, M. K., A. Sele, and K. M. Lund. 2000. Design of a 25.2" titanium catenary riser for the Asgard B platform. Paper OTC 12030, presented at the Offshore Technology Conference, Houston;
Slagsvold, L., and G. Langford. 2000. *Composite Catenary Production Riser.* New Orleans: DOT.

[4]Edwards, R. Y., R.Z. Filho, W.F. Hennessy, C.R. Campman, H. Bailon. 1999. Load monitoring at the touch down point of the first steel catenary riser installed in a deepwater moored semisubmersible platform. Paper OTC 10975, presented at the Offshore Technology Conference, Houston;
Croutelle, Y., C. Ricbourg, C. Le Cunff, J.-M. Heurtier, and F. Biolley. 2002. Fatigue of steel catenary risers. Paper presented at the Ultra Deep Engineering and Technology Conference, Brest, France;
Bridge, C., and N. Willis. 2002. Steel catenary risers—results and conclusions from large scale simulations of seabed interaction. Paper presented at the Deep Oil Technology Conference;
Lee, L., and D. Allen. 2004. VIV modeling of bare and suppressed risers. Paper OTC 16183, presented at the Offshore Technology Conference, Houston.

13

Axial Vibrations of Fixed Risers

There are similarities between the axial vibrations of *fixed* risers (which are the subject of this chapter), the axial vibrations of *hung-off* risers, and the transverse vibrations of fixed risers (which are treated in the following chapters). Since the axial vibrations of fixed risers are the simplest to study, this chapter serves as a general introduction to riser vibrations.

Axial vibrations of fixed risers are rarely a problem in the real world, but they were pertinent to an interesting project undertaken by the Ocean Drilling Program (ODP) in the 1990s. As part of their program of scientific coring of the seabed, the ODP attempted to adapt a diamond coring system, as used in the mining industry, to coring below the deep ocean.[1] Axial vibrations were one of the problems they encountered in that attempt.

The ODP Experience

ODP attempted to build a *piggyback* coring system, for use with the drill-ship *Joides Resolution* in ultradeep water (4,500 m). The plan was to use the ship's drill string, anchored to the seabed, as a *mini-riser* to guide a small diameter mining string, equipped with a diamond coring bit. The mini-riser was tensioned by the ship's main (800 kip) compensator, and the idea was to work the mining string from a platform installed at the top end of that mini-riser. Such systems work well for geotechnical drilling in modest water depths (up to 500 m), but the weight on bit (WOB) has to be very carefully controlled, to avoid destruction of the diamond bit.

For ODP, the greatly increased water depth and the inevitable fluctuations in tension applied by the large-capacity compensator caused major problems in the form of fluctuating stretch of the mini-riser. The top end of the mini-riser could not be used directly as a *fixed reference point* for the mining string. Attempts to do so led to unacceptable fluctuations of WOB that led to the destruction of the diamond bit. ODP chose to install a *secondary heave compensator* between the top ends of the mining string and the mini-riser and to *pilot* the secondary compensator to adjust for the axial stretch of the mini-riser.

Attempts were made to deduce the fluctuating stretch of the mini-riser by measuring the fluctuating tension at the top end. This is difficult, even for a pure sinusoidal excitation, since the relation between dynamic tension and stretch depends on the frequency. With the available technology, it was finally not found possible to deduce the fluctuating stretch with sufficient precision. The stretch was far from being a pure sinusoid because of stick-slip friction in the main compensator. The project was finally abandoned. The relation between dynamic tension and stretch, which was the heart of the problem, becomes clear when the basic physics of the problem is analyzed as in the following sections.

Axial Stress Waves

Figure 13–1*a* and *b* show a vertical *uniform* riser and a vertical *segmented* riser, respectively, subject to a top-end sinusoidal movement of amplitude U_0 and frequency ω, where u is the dynamic vertical displacement and, hence, du/dx is the dynamic strain.

Figure 13–1*c* shows the internal dynamic axial forces acting on a short element of length δx and mass m per unit length. The dynamic axial force is related to the local strain by

$$T = EA\frac{\partial u}{\partial x} \qquad (13.1)$$

where E is the riser material modulus and A is the cross-sectional area. Since the mass-acceleration of the element is equal to the applied force, from figure 13–1*c*,

$$m\,\delta x\frac{\partial^2 u}{\partial t^2} = \delta T = \delta\left(EA\frac{\partial u}{\partial x}\right) \qquad (13.2)$$

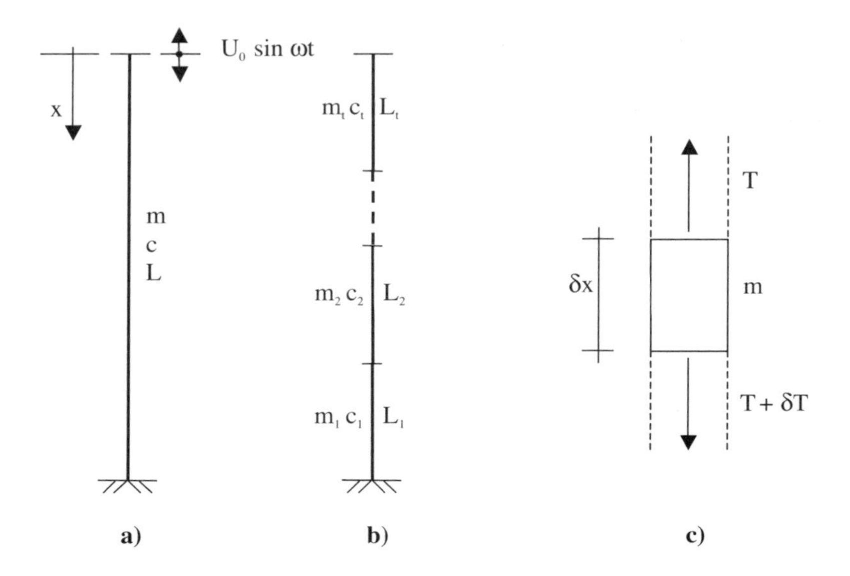

Fig. 13–1. *Riser axial vibrations—symbols and internal forces*

Hence,

$$m\frac{\partial^2 u}{\partial t^2} = EA\frac{\partial^2 u}{\partial x^2} \qquad (13.3)$$

Equation (13.3) is the *wave equation,* which can be written as

$$\frac{\partial^2 u}{\partial t^2} = c^2\frac{\partial^2 u}{\partial x^2} \qquad (13.4)$$

where c is the *celerity* (speed of transmission of axial stress waves in the riser). From equations (13.4) and (13.4), $mc^2 = EA$. The celerity c is therefore given by

$$c = \sqrt{\frac{EA}{m}} \qquad (13.5)$$

Note that if the riser mass is entirely *structural* (with no additional mass in the form of buoyancy modules or peripheral lines, etc.), then $m = \rho A$, and $c = \sqrt{E/\rho}$, where ρ is the mass density of the riser material.

Axial Displacement-Tension Relationships for a Uniform Riser

For the configuration shown in figure 13–1*a,* the vertical displacement at distance x below the top end is given by the solution to equation (13.4), which yields

$$u_{x,t} = U_0 \frac{\sin[\omega(L-x)/c]}{\sin(\omega L/c)} \sin \omega t \qquad (13.6)$$

The dynamic tension is related to the local strain by equation (13.1). Hence, it can be found by differentiating equation (13.6). Since $EA = mc^2$, the dynamic tension is given by

$$T_{x,t} = -mc\omega U_0 \frac{\cos[\omega(L-x)/c]}{\sin(\omega L/c)} \sin \omega t \qquad (13.7)$$

The amplitude of the top-end dynamic tension T_{top} is given by equation (13.8) (for $x = 0$), which depends on the frequency ω:

$$T_{top} = -\frac{m\omega c}{\tan(\omega L/c)} U_0 \qquad (13.8)$$

The *Amplitude* work sheet of the Excel file *Fixed-Axial-Vibrations.xls* shows graphically the amplitudes of the axial displacements U_x and the dynamic tensions T_x for a unit amplitude top-end displacement (±1 m), for different riser data and excitation period. The work sheet also gives the number n of the closest resonant mode and the corresponding value of the period T_{pn}.

Responses of a Uniform Riser, Using Excel File *Fixed-Axial-Vibrations.xls*

Axial resonance occurs when the denominators of equations (13.6) and (13.7) are zero—namely, for frequencies ω_n given by equation (13.9) for which sin $(\omega_n L/c) = 0$, where n is the mode number:

$$\frac{\omega_n L}{c} = n\pi \qquad (13.9)$$

Since $\omega_n = 2\pi/T_{pn}$, where T_{pn} is the period of the vibration for mode n, resonance occurs for natural periods T_{pn} given by

$$T_{pn} = \frac{2L}{nc} \qquad (13.10)$$

The *Mode* work sheet of *Fixed-Axial-Vibrations.xls* gives the natural periods T_{pn} and corresponding values of the frequency ω_n for the first 20 modes, according to the input riser data. If any of these values is used as input with the *Amplitude* work sheet, the responses are infinite but give the *mode shapes*.

Equation (13.10) can be rewritten as follows:

$$T_{pn} = 2\left(\frac{L_n}{c}\right) \qquad (13.11)$$

where L_n is the internodal distance ($L_n = L/n$). Hence, at resonance, the period is always equal to *twice the time required for a stress wave to run between adjacent nodes*. This can be clearly seen in figure 13–2, in which the modal responses of the displacements are shown (traced horizontally) for a 2,000 m riser for modes 1, 2, and 4. If, for example, the riser celerity is 2,000 m/s, equations (13.10) and (13.11) both give natural periods of 2, 1, and 0.5 s for those respective modes.

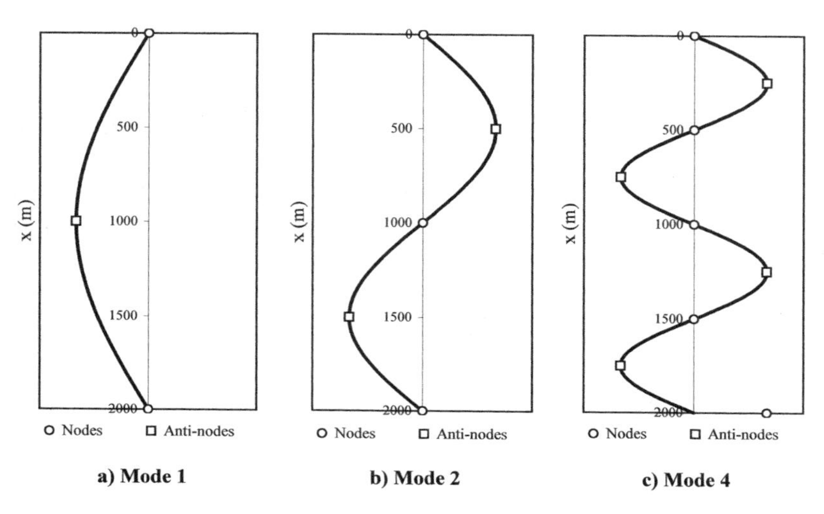

Fig. 13–2. *Example modal displacement responses for a uniform 2,000 m riser (vertical displacements traced horizontally)*

Figure 13–3 gives the modal responses for the dynamic tensions for the same riser. As can be seen, they are 90° out of phase with the vertical displacements, as expected from equations (13.6) and (13.7). Hence, maximum tension fluctuations occur at the *nodes,* where the displacements are zero. Likewise, dynamic tensions are zero at the *antinodes,* where the displacements are maximum.

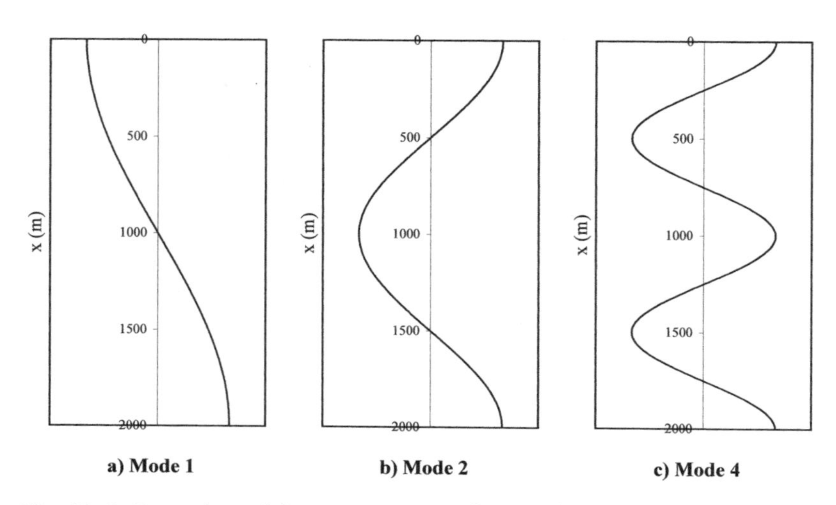

Fig. 13–3. *Example modal tension responses for a uniform 2,000 m riser (dynamic tensions traced horizontally)*

Equation (13.6) can be rewritten as equation (13.12), which is the sum of two waves:

$$u_{x,t} = \frac{U_0}{2\sin(\omega L/c)}\left[-\cos\omega\left(t+\frac{L-x}{c}\right)+\cos\omega\left(t-\frac{L-x}{c}\right)\right] \qquad (13.12)$$

In equation (13.12), the first term within brackets is the descending wave, and the second is the ascending wave. Note that since the displacement is zero at the lower end ($x = L$), reflection there causes a change in sign of the amplitudes of the two waves.

The dynamic tension given by equation (13.7) can also be rewritten as the sum of descending and ascending waves, as follows:

$$T_{x,t} = \frac{-mc\omega U_0}{2\sin(\omega L/c)}\left[\sin\omega\left(t+\frac{L-x}{c}\right)+\sin\omega\left(t-\frac{L-x}{c}\right)\right] \qquad (13.13)$$

The *Waves* work sheet of *Fixed-Axial-Vibrations.xls* shows the descending and ascending waves of equations (13.12) and (13.13). It allows the reader to observe how the waves progress and combine to give the riser displacement and dynamic tension responses. The work sheet allows the reader to advance the time by intervals equal to 1/32nd of the excitation period.

Dynamic Stiffness of a Uniform Riser

The dynamic stiffness is given by the ratio of the amplitude of the dynamic tension to the amplitude of the top-end movement T_x/U_0. Again, $mc^2 = EA$, so from equation (13.7), the dynamic stiffnesses at the top end and bottom end of a uniform riser are given by equations (13.14) and (13.15), respectively:

$$\frac{T_{\text{top}}}{U_0} = \frac{EA}{L}\left[\left(\frac{\omega L}{c}\right)\Big/\tan\left(\frac{\omega L}{c}\right)\right] \qquad (13.14)$$

$$\frac{T_{\text{btm}}}{U_0} = \frac{EA}{L}\left[\left(\frac{\omega L}{c}\right)\Big/\sin\left(\frac{\omega L}{c}\right)\right] \qquad (13.15)$$

where T_{top} and T_{btm} are the amplitudes of the dynamic tensions at the top and bottom ends. The bracketed terms in equations (13.14) and (13.15) are the *dynamic stiffness factors*. As expected and required, for low frequencies, both factors tend toward 1.

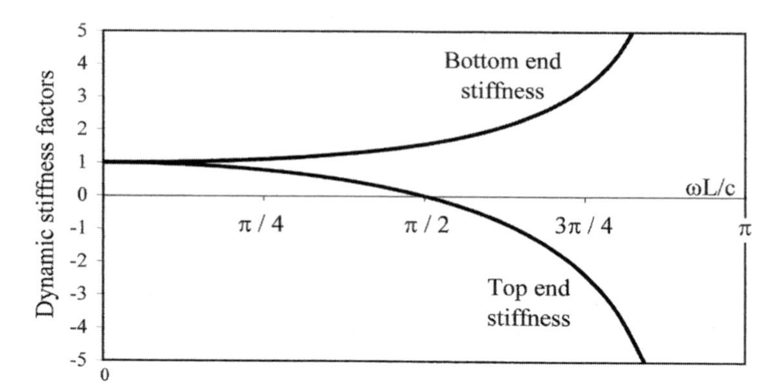

Fig. 13–4. *Top- and bottom-end dynamic stiffness factors*

The dynamic stiffness factors of equations (13.14) and (13.15) are plotted in figure 13–4 for the range $0 < \omega L/c < \pi$, which corresponds to periods longer than the fundamental. The ratio of top- to bottom-end stiffness factors is equal to $\cos(\omega L/c)$. The amplitudes of the two responses are equal only at resonant frequencies ($\omega_n L/c = n\pi$), for which the responses are infinite. For all nonresonant frequencies ($-1 < \cos(\omega L/c) < 1$), the amplitude of the dynamic force is always *greater at the bottom end* than at the top end.

From equation (13.14), for $\omega L/c = \pi/2$, $3\pi/2$, $5\pi/2$, and so forth, the denominator is infinite. However, those frequencies correspond to the following multiples of the fundamental period: 2, 2/3, 2/5, and so forth. Hence, at those periods, the top-end stiffness is zero. These results can be confirmed using the *Amplitude* work sheet of *Fixed-Axial-Vibrations.xls*. The zero top-end stiffness at twice the fundamental period (for $\omega L/c = \pi/2$) can be clearly seen in figure 13–4.

Axial Vibration of a Segmented Riser

For a segmented riser, as shown in figure 13–1*b,* a similar approach can be used. The segments are numbered from the lower end upward (subscripts "*s*," "*s* + 1," etc.), and local coordinates are used for each segment, with *x* measured from the segment top end. For each segment, the displacements and dynamic tensions can be expressed in terms of the amplitude U_0 at the segment top end and an equivalent length L', both of which have to be determined.

The bottom-end displacement of segment $s + 1$ is equal to the top-end displacement of segment s. Hence, from equation (13.6),

$$\left[U_0 \frac{\sin \frac{\omega(L'-L)}{c}}{\sin \frac{\omega L'}{c}} \right]_{s+1} = [U_0]_s \qquad (13.16)$$

Likewise, equating the bottom-end dynamic tension of segment $s + 1$ to the top-end dynamic tension of segment s gives, from equation (13.7),

$$\left[mc\omega U_0 \frac{\cos \frac{\omega(L'-L)}{c}}{\sin \frac{\omega L'}{c}} \right]_{s+1} = \left[mc\omega U_0 \frac{\cos \frac{\omega L'}{c}}{\sin \frac{\omega L'}{c}} \right]_s \qquad (13.17)$$

Hence, dividing equation (13.16) by (13.17) yields

$$\left[\left(\frac{1}{mc} \right) \tan \frac{\omega(L'-L)}{c} \right]_{s+1} = \left[\left(\frac{1}{mc} \right) \tan \frac{\omega L'}{c} \right]_s \qquad (13.18)$$

In equations (13.16) – (13.18), all the bracketed characteristics (U_0, m, c, L', and L) refer to the particular segment.

For the lowest segment ($s = 1$), $L' = L$, which is the true length of the segment. For each of the other segments, L' is an equivalent length. Values of L' can be found progressively, by working up from the riser bottom end. The relative amplitudes $[U_0]$ of the displacements at the junctions between segments can then be determined from equation (13.16).

By use of the subscript "t" to denote the segment at the top end, resonance will occur for $[\omega_n L'/c]_t = n\pi$ (cf. equation [13.9]). The resonant values of ω_n can be found by iteration. Caution is required when using equation (13.18) to find the resonant frequencies. At each junction between segments, it is important to conserve the same number of integral multiples of π for $[\omega_n(L' - L)/c]_{s+1}$ as for $[\omega_n L'/c]_s$.

Resonance again occurs when the returning axial wave is in phase with the top-end excitation. Hence, the fundamental resonant period T_{pl} is again equal to the time for a stress wave to descend and reascend the riser of equivalent length L'_{t}, as given by $T_{\text{pl}} = 2\,L'_t \,/\, c_t$.

Summary

The axial stress wave equation has been derived, and the axial displacement and dynamic tension equations have been deduced for fixed uniform risers. Dynamic tension fluctuations have been shown to depend on the frequency and the amplitude of the top-end excitation.

Resonant periods are shown to be equal to twice the time required for a stress wave to run between adjacent nodes. The fundamental resonant period is equal to the time taken for a stress wave to descend and reascend the riser.

The dynamic stiffness (ratio of amplitude of dynamic tension to amplitude of top-end displacement) has been shown to be always greater at the riser bottom-end than at the top end at nonresonant periods. At twice the fundamental natural period, the top-end dynamic stiffness is zero.

For uniform risers, the Excel file *Fixed-Axial-Vibrations.xls* can be used to confirm the results obtained in this chapter. The procedure for calculating the responses and the resonant frequencies of a segmented riser has also been explained.

Axial resonance of hung-off risers can now be examined in the next chapter.

References

[1]Storms, M., S. Howard, D. Reudelheuber, G. Holloway, P. Rabinowitz, and B. Harding. 1990. A slimhole coring system for the deep oceans. SPE/IADC 21907;

Sparks, C. P. 1995. Riser technology for deepwater scientific drilling in the 21st century. Paper presented at the 14th Offshore Mechanics and Arctic Engineering Conference, Copenhagen.

14

Axial Vibrations of Hung-Off Risers

Axial vibration is of particular concern for drilling risers hung off drilling vessels in storm conditions. In such conditions, hung-off risers are subject to axial excitation induced by the heave of the drilling vessel. The danger is that unacceptably large axial forces may be induced in the riser, which may even lead to dynamic buckling. The subject has been treated in a number of publications in recent years.[1]

Analysis of axial vibrations of hung-off risers is more complicated than for fixed risers for several reasons. First, the risers generally have a large concentrated mass at the lower end in the form of a lower marine riser package (LMRP) or blowout preventer (BOP). Second, the resonant response depends on short-period heave of the drilling vessel and on the riser axial damping, both of which are difficult to determine precisely.

Nevertheless, a similar approach to the previous chapter can be used to determine resonance frequencies and to understand the phenomenon. Figure 14–1 shows six riser models that can be studied analytically.

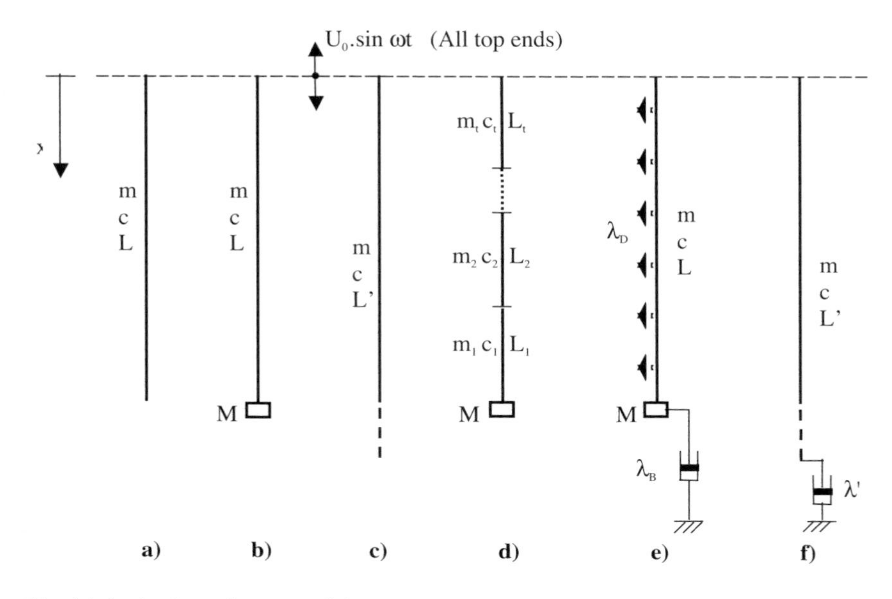

Fig. 14–1. *Analytical riser models*

Uniform Riser

Figure 14–1*a* shows a uniform riser, which can be analyzed very simply by using equations similar to those of the previous chapter. The tension T_L at the riser bottom end is always zero. Hence, from the tension-strain relationship (see equation [13.1]), $T_L = EA(\partial u/\partial x)_L = 0$. Again, if it is noted that $EA = mc^2$ (see equation [13.5]), where m is the mass per unit length and c is the celerity (or speed of transmission of axial stress waves), then the wave equation (see equation [13.4]) yields the following solutions for the displacement $u_{x,t}$ and the dynamic tension $T_{x,t}$ at distance x below the top end (cf. eqq. [13.6] and [13.7]):

$$u_{x,t} = U_0 \frac{\cos[\omega(L-x)/c]}{\cos(\omega L/c)} \sin\omega t \qquad (14.1)$$

and

$$T_{x,t} = mc\omega U_0 \frac{\sin[\omega(L-x)/c]}{\cos(\omega L/c)} \sin\omega t \qquad (14.2)$$

Resonance occurs for $\cos(\omega_n L/c) = 0$, for which $\omega_n L/c = (2n - 1)\pi/2$. Since the period $T_{pn} = 2\pi/\omega_n$, resonance occurs at periods given by

$$T_{pn} = \frac{4L}{(2n-1)c} \qquad (14.3)$$

where n is the mode number and $n = 1$ denotes the fundamental. The fundamental resonant period is therefore equal to $4L/c$, which is the time taken by an axial stress wave to run four times the length of the riser.

Uniform Riser with Concentrated Mass at Lower End

For a riser with a concentrated mass M at the lower end, in the form of an LMRP or a BOP, the analysis is more complicated. The wave equation (see equation [13.4]) is satisfied by

$$u_{x,t} = U_0 \left(\cos\frac{\omega x}{c} + B_0 \sin\frac{\omega x}{c} \right) \sin \omega t \qquad (14.4)$$

where B_0 is a constant that depends on the concentrated mass at the lower end. The constant B_0 can be determined by considering the forces that act on the concentrated mass at the lower end, as given by

$$mc^2 \left(\frac{\partial u}{\partial x} \right)_{L,t} = -M \left(\frac{\partial^2 u}{\partial t^2} \right)_{L,t} \qquad (14.5)$$

where the left-hand side of the equation is the force resulting from the riser dynamic strain, and the right-hand side is the inertia force of the concentrated mass. Substitution of equation (14.4) into equation (14.5) then leads to the value of B_0:

$$B_0 = \frac{M\omega + mc\tan(\omega L/c)}{mc - M\omega\tan(\omega L/c)} \qquad (14.6)$$

The constant B_0 can be expressed in terms of a new constant L' defined by:

$$B_0 = \tan\left(\frac{\omega L'}{c}\right) \tag{14.7}$$

Substitution for B_0 in equation (14.6) then leads to

$$mc\tan\frac{\omega(L'-L)}{c} = M\omega \tag{14.8}$$

Substitution of B_0 into equation (14.4) leads to equation (14.9) for the axial displacement at distance x from the top end:

$$u_{x,t} = U_0\frac{\cos[\omega(L'-x)/c]}{\cos(\omega L'/c)}\sin\omega t \tag{14.9}$$

Again, note that $EA = mc^2$; thus, the dynamic tension is given by equation (14.10), since $T_{xt} = mc^2(\partial u/\partial x)$:

$$T_{x,t} = mc\omega U_0\frac{\sin[\omega(L'-x)/c]}{\cos(\omega L'/c)}\sin\omega t \tag{14.10}$$

By comparison of equations (14.1) and (14.9), it can be seen that L' is an equivalent length of uniform riser, as shown in figure 14–1*c*. Note that for low frequencies, equation (14.8) tends toward equation (14.11). The riser behaves as though its length were extended by M/m. Equation (14.11) gives the maximum value of L'.

$$L' = L + \frac{M}{m} \tag{14.11}$$

As the frequency ω *increases,* the precise value of L' is *reduced.*

From Equations 14.9 and 14.10, resonance occurs for $\cos(\omega_n L'/c) = 0$ —and, hence, for values of $\omega_n L'/c$ given by

$$\frac{\omega_n L'}{c} = (2n-1)\frac{\pi}{2} \tag{14.12}$$

where n is the mode number.

Equations (14.8) and (14.12) can be combined to give equation (14.13), in which the first equality is a rewritten version of equation (14.8):

$$\frac{\omega_n L'}{c} = \frac{\omega_n L}{c} + \tan^{-1}\left(\frac{M\omega_n}{mc}\right) = (2n-1)\frac{\pi}{2} \tag{14.13}$$

Good approximations for the resonant values of ω_n can be found by substituting the approximate value of L' from equation (14.11) into equation (14.12). The precise value of ω_n, for a particular mode, can then be quickly found by iteration, using the second equality of equation (14.13). The corresponding value of L' can then be found by substituting the final values of ω_n back into equation (14.12).

Since the period $T_{pn} = 2\pi/\omega_n$, the resonance periods T_{pn} are as follows (cf. equation [14.3]):

$$T_{pn} = \frac{4L'}{(2n-1)c} \tag{14.14}$$

where n is the mode number ($n = 1$ denotes the fundamental). Hence, the fundamental resonant period T_{p1} is equal to the time taken by an axial stress wave to run four times the riser equivalent length L'.

Simulations Using Excel File *Hungoff-Free-Vibrations.xls*

The *Amplitude* work sheet of the Excel file *Hungoff-Free-Vibrations.xls* shows graphically the amplitude ratios of the axial displacements U_x/U_0 and the dynamic tensions $T_x/mc\omega U_0$, given by equations (14.9) and (14.10), for different data. Table 14–1 gives example data used to generate the responses shown in figure 14–2 for modes 1, 2, and 4. The modal periods, frequencies, and equivalent lengths L' are also given in table 14–1.

Table 14–1. *Example data used for figure 14–2*

m (tonnes/m)		0.6	
c (m/s)		2,500	
L (m)		2,100	
M (tonnes)		180	
Mode number	1	2	4
Period T_p (s)	3.83	1.26	0.53
ω (radians/s)	1.639	4.970	11.945
L' (m)	2,396	2,371	2,301

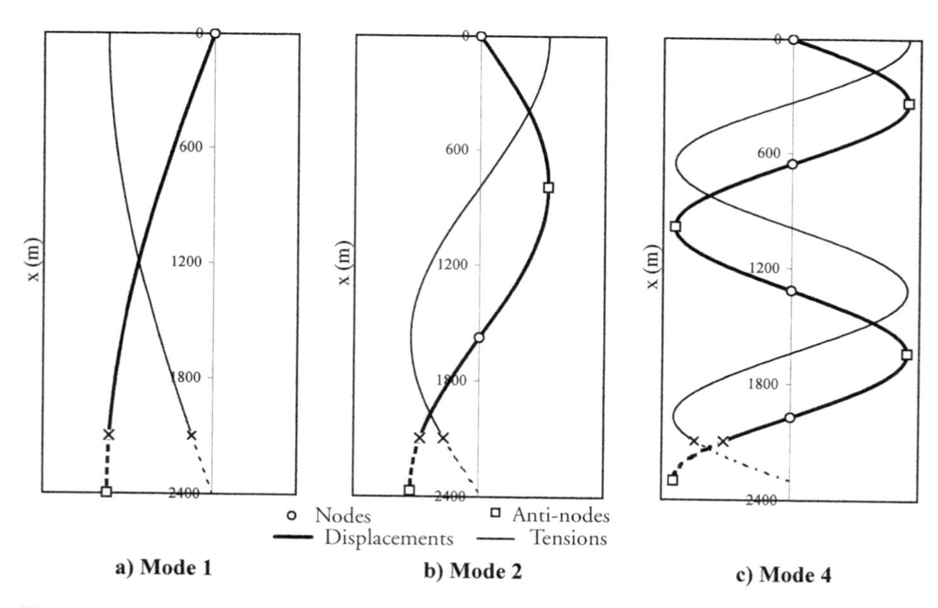

Fig. 14–2. *Example modal displacement and dynamic tension responses*

The extra equivalent lengths ($L' - L$) are depicted as dotted lines in figure 14–2. As can be seen from table 14–1, the equivalent lengths reduce as the mode number increases. Equation (14.11) gives an approximate value of $L' = 2{,}400$ m.

The maximum dynamic tensions are seen to occur at the nodes. This is already clear by comparison of equations (14.1) and (14.2). Likewise, dynamic tensions are zero at the antinodes, where the axial displacements are maximum. *Hungoff-Free-Vibrations.xls* also gives the number n of the closest resonant mode and the corresponding resonant period T_{pn}. The *Mode* work sheet gives the resonant periods T_{pn} and corresponding frequencies ω_n and equivalent lengths L', for the first 20 modes and for mode numbers specified by the reader.

The responses given by equations (14.9) and (14.10) can be represented as the sum of ascending and descending waves, as in chapter 13 (cf. eqq. [13.12] and [13.13]). They are then given by equations (14.15) and (14.16), respectively:

$$u_{x,t} = \frac{U_0}{2\cos(\omega L'/c)}\left[\sin\omega\left(t+\frac{L'-x}{c}\right)+\sin\omega\left(t-\frac{L'-x}{c}\right)\right] \qquad (14.15)$$

$$T_{x,t} = \frac{mc\omega U_0}{2\cos(\omega L'/c)}\left[-\cos\omega\left(t+\frac{L'-x}{c}\right)+\cos\omega\left(t-\frac{L'-x}{c}\right)\right] \qquad (14.16)$$

The *Waves* work sheet of *Hungoff-Free-Vibrations.xls* shows the descending and ascending waves of equations (14.15) and (14.16) and how they combine to give the riser responses. The work sheet allows the reader to advance the time t by steps equal to 1/32nd of the excitation period.

Segmented Riser with Concentrated Mass at Lower End

For a segmented riser, as shown in figure 14–1*d,* an approach similar to that of chapter 13 can be used. The segments are again numbered from the lower end upward (by subscripts "s," "s + 1," etc.), and local coordinates are used for each segment, with x measured *downward* from the segment top end. The displacements and the dynamic tensions can be expressed in terms of the amplitude U_0 of the *segment* top end displacement and an equivalent length L', both of which have to be determined for each segment.

Equating the amplitude of the bottom-end displacement of segment $s + 1$ to the amplitude of the top-end displacement of segment s leads to the following expression (see also equation [14.9]):

$$\left[U_0 \frac{\cos\dfrac{\omega(L'-L)}{c}}{\cos\dfrac{\omega L'}{c}}\right]_{s+1} = [U_0]_s \qquad (14.17)$$

Likewise, equating the amplitude of the bottom end dynamic tension of segment $s + 1$ to the amplitude of the top end dynamic tension of segment s leads to the following expression (see also equation [14.10]):

$$\left[mc\omega U_0 \frac{\sin\dfrac{\omega(L'-L)}{c}}{\cos\dfrac{\omega L'}{c}}\right]_{s+1} = \left[mc\omega U_0 \frac{\sin\dfrac{\omega L'}{c}}{\cos\dfrac{\omega L'}{c}}\right]_s \qquad (14.18)$$

Dividing equation (14.18) by equation (14.17) yields

$$\left[mc\tan\frac{\omega(L'-L)}{c}\right]_{s+1} = \left[mc\tan\frac{\omega L'}{c}\right]_s \qquad (14.19)$$

In equations (14.17) – (14.19), all the bracketed characteristics (U_0, m, c, L', and L) refer to the particular segment. Equation (14.19) allows all values of L' to be found, by working up from the riser bottom end. For the first (lowest) segment, equation (14.19) has to be replaced by equation (14.8). The amplitudes U_0 of the displacements at the junctions between segments can then be found from equation (14.17).

Resonance will occur when for the uppermost segment $[\omega_n L'/c]_t = (2n-1)\pi/2$. The corresponding values of the frequency ω_n can be found by iteration. Since $T_p = 2\pi/\omega$, the resonant periods T_{pn} are given by the following expression (cf. equations [14.3] and [14.13]):

$$T_{pn} = \frac{1}{2n-1}\left(\frac{4L'}{c}\right)_t \qquad (14.20)$$

where the subscript "t" refers to the segment at the riser top end, and n is the mode number. The fundamental period T_{p1} is again equal to the time required for a stress wave to run four times the equivalent length of the riser, calculated for the top-end segment.

Caution is required when using equation (14.19) to find resonant frequencies. The same number of integral multiples of π must be conserved for $[\omega(L'-L)/c]_{s+1}$ as for $(\omega L'/c)_s$.

Riser Comprising Multiple Repeated Joints

Real risers consist of long segments composed of large numbers of identical joints comprising riser tubes and connectors. The reader may wonder whether a more accurate evaluation of the axial response could be obtained by modeling every joint, particularly if the tubes and connectors are made of different materials. Such detailed modeling is not necessary, as can be seen by comparing results for a particular case. Table 14-2 gives data for a riser composed of 80 joints, 23 m long, excited at a period of 0.65 s, which is close to the third-mode resonant period (0.63 s). Figure 14–3 shows the resulting dynamic tensions.

Table 14–2. *Data for example calculation of dynamic tension ratios*

Joint characteristics	Tube	Connector
Length, L (m)	22	1
Mass per unit length, m (tonnes/m)	1.1	2.3
Axial stiffness, EA (kN)	7,000,000	49,000,000
Celerity, c (m/s)	2,558	4,636
Riser characteristics		
Mass of LMRP, M (tonnes)	200	
No. of joints	80	
Total length (m)	1,840	
Total mass, without LMRP (tonnes)	2,120	
Excitation and equivalent parameters		
Excitation period, T_p (s)	0.65	
Mean mass, m (tonnes/m)	1.15	
Mean stiffness, EA (kN)	7,270,968	
Mean celerity, c (m/s)	2,512	
Equivalent length, L' (m)	1,993	

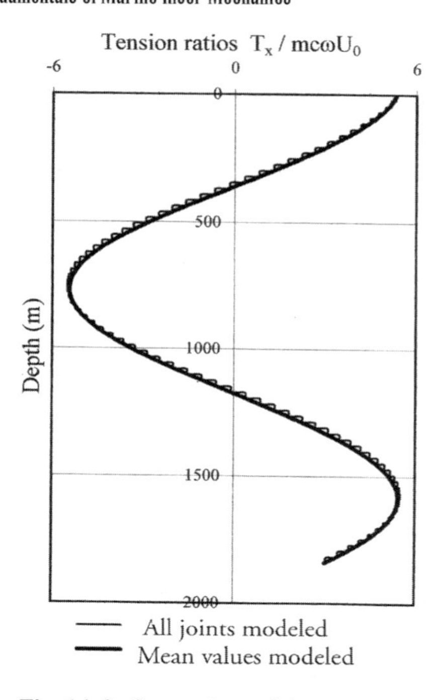

Fig. 14–3. *Comparison of dynamic tensions induced in two riser models*

As figure 14–3 illustrates, the model using mean values gives virtually the same dynamic tensions as those obtained by modeling all joints. The small fluctuations about the mean for the model in which all joints are simulated can be clearly seen.

It may appear obvious to the reader that it is sufficient to model mean values. However, it is less obvious when the propagation of axial stress waves is considered. The good agreement results from the fact that the axial stress waves are partially reflected at the junctions of tubes and connectors. The reflected waves combine with—and modify—the returning wave. As a result, the returning wave progresses with a *mean celerity* equal to the celerity calculated from mean values.

For each segment (consisting of any number of identical joints), it is therefore sufficient to use the mean values of mass per unit length m and celerity c. The mean mass per unit length m is the total mass of the segment divided by its length. The mean celerity c must be calculated from $c = \sqrt{EA/m}$ using the segment mean axial stiffness EA_{mean}. The latter is obtained from the segment stretch induced by an axial force. If an axial force F induces a stretch e in a segment of length L, then $EA_{mean} = FL/e$.

Uniform Riser with Distributed Damping

Damping can be taken into account analytically providing that it is linear or linearized. Figure 14–1*e* shows a uniform riser with linear damping along its length and at the LMRP level. Again, note that EA = mc²; thus, the *damped-wave equation* for the riser is given by

$$m\frac{\partial^2 u}{\partial t^2}=mc^2\frac{\partial^2 u}{\partial x^2}-\lambda_D\frac{\partial u}{\partial t} \tag{14.21}$$

The top-end condition is satisfied by the imaginary part of the following expression:

$$u_x=U_0\left[B_1e^{-jKx}+(1-B_1)e^{jKx}\right]e^{j\omega t} \tag{14.22}$$

where B_1 and K are complex numbers.

Substitution of equation (14.22) into (14.21) gives the value of K (i.e., $K = K_1 - jK_2$). The equilibrium of forces acting on the mass M at the riser bottom end ($x = L$) requires that

$$M\left(\frac{\partial^2 u}{\partial t^2}\right)_L+mc^2\left(\frac{\partial u}{\partial x}\right)_L+\lambda_B\left(\frac{\partial u}{\partial t}\right)_L=0 \tag{14.23}$$

After much working, equations (14.22) and (14.23) give the value of B_1. Extracting the imaginary part of equation 14.22 then yields the expressions for the displacements and dynamic tensions as a function of x. These expressions are given in appendix I. The amplitudes of the displacements U_x and the dynamic tensions T_x are finally given by equations (14.24) and (14.25):

$$U_x=U_0\sqrt{R_x-S_x} \tag{14.24}$$

$$T_x=mc\omega U_0\sqrt{\frac{R_x+S_x}{\cos 2\phi}} \tag{14.25}$$

where the functions R_x and S_x vary with x and are given by equations (14.26) and (14.27):

$$R_x = \overline{1+\alpha}\left(e^{2K_2x}\right)+\left(\frac{\alpha^2+\beta^2}{2}\right)\cosh 2K_2x \qquad (14.26)$$

$$S_x = \left(\frac{\alpha^2+\beta^2}{2}+\alpha\right)\cos 2K_1x-\beta\sin 2K_1x \qquad (14.27)$$

Expressions for the symbols α, β, K_1, K_2, and ϕ are given in appendix I in terms of a chain of further parameters.

Uniform Riser with Equivalent Damping

The displacements and dynamic tensions of the equivalent riser are simple to express if damping is considered to be concentrated at the lower extremity of the equivalent length, called the *equivalent end* (at $x = L'$) as modeled in figure 14–1*f*. At first sight, this model may appear to be too far from reality to be of any interest. However, if the equivalent damping factor is chosen carefully, it gives results that are generally in excellent agreement with those obtained earlier, as will be seen in the next section. Therefore, it can help in understanding riser vibrational behavior.

The wave equation (equation [13.2]) and the top-end imposed displacement (at $x = 0$) are satisfied by

$$u_{x,t} = U_0\left[\left(B_2 \sin\frac{\omega x}{c}+\cos\frac{\omega x}{c}\right)\sin\omega t + B_3 \sin\frac{\omega x}{c}\cos\omega t\right] \qquad (14.28)$$

where B_2 and B_3 are constants to be determined. If a damping coefficient (λ') is applied at the lower equivalent end (at $x = L'$), then consideration of equilibrium there leads to equation (14.29), which allows the constants B_2 and B_3 to be found:

$$\lambda'\left(\frac{\partial u}{\partial t}\right)_{L'} + mc^2\left(\frac{\partial u}{\partial x}\right)_{L'} = 0 \qquad (14.29)$$

The solution is given in appendix I. The resultant amplitudes U_x and T_x of the displacement and the dynamic tension at x are then given by equations (14.30) and (14.31):

$$U_x = U_0 \sqrt{\frac{\cos^2 \dfrac{\omega(L'-x)}{c} + \gamma^2 \sin^2 \dfrac{\omega(L'-x)}{c}}{\cos^2 \dfrac{\omega L'}{c} + \gamma^2 \sin^2 \dfrac{\omega L'}{c}}} \qquad (14.30)$$

$$T_x = mc\omega U_0 \sqrt{\frac{\sin^2 \dfrac{\omega(L'-x)}{c} + \gamma^2 \cos^2 \dfrac{\omega(L'-x)}{c}}{\cos^2 \dfrac{\omega L'}{c} + \gamma^2 \sin^2 \dfrac{\omega L'}{c}}} \qquad (14.31)$$

where $\gamma = \lambda'/mc$. Note in both equations (14.30) and (14.31), the maximum value of the numerator, below the square root sign, is equal to 1. Note also that, at resonance, $\cos^2(\omega L'/c) = 0$, and $\sin^2(\omega L'/c) = 1$. Hence, at resonance, the two equations yield maximum displacements and dynamic tensions given by: $U_{max}/U_0 = T_{max}/mc\omega U_0 = 1/\gamma$.

Simulations Using Excel File *Hungoff-Damped-Vibrations.xls*

The Excel file *Hungoff-Damped-Vibrations.xls* can be used to compare the amplitudes of the displacements and dynamic tensions calculated using equations (14.24) and (14.25) with those given by equations (14.30) and (14.31). For the comparison to have significance, the equivalent damping coefficient λ' (and, hence, γ [since $\gamma = \lambda'/mc$]) at the *equivalent lower end* must be chosen to give the same mean rate of dissipation of energy as the combined effect of the damping coefficients λ_D and λ_B of the riser with distributed damping.

As shown in appendix I, the required value of γ is given by

$$\frac{\gamma}{\cos^2(\omega L'/c) + \gamma^2 \sin^2(\omega L'/c)} = \frac{\lambda_D}{mc}\int_0^L (R_x - S_x)\,dx + \frac{\lambda_B}{mc}(R_L - S_L) \qquad (14.32)$$

where R_L and S_L are the values of R_x and S_x at $x = L$. At resonant frequencies, $\cos^2(\omega L'/c) = 0$, and $\sin^2(\omega L'/c) = 1$; hence, equation (14.32) reduces to

$$\frac{1}{\gamma} = \frac{\lambda_D}{mc}\int_0^L (R_x - S_x)\,dx + \frac{\lambda_B}{mc}(R_L - S_L) \qquad (14.33)$$

Hungoff-Damped-Vibrations.xls allows the responses of the riser with distributed damping (fig. 14–1*e*) to be compared to those of the riser with damping at the equivalent lower end (fig. 14–1*f*) for different data. *Hungoff-Damped-Vibrations.xls* includes two work sheets: *Forced-Vibrations,* for which the period T_p of the top-end excitation must be specified; and *Modal-Vibrations,* which allows the vibration mode number to be specified. For both work sheets, the data are input for the model of figure 14–1*e,* consisting of the values of m, c, L, M, λ_B, and λ_D.

The resulting amplitudes of the displacements and dynamic tensions are presented graphically. The maximum amplitude ratios U_{max}/U_0 and $T_{max}/mc\omega U_0$ are compared for the two models. The top-end amplification factors $T_{top}/U_0\omega^2\Sigma M$ are also compared for the two models, where ΣM is the riser total mass.

Agreement between the results for the two models is generally excellent, unless the distributed damping λ_D is very large, in which case the riser model of figure 14–1*e* behaves as though it were being pumped up and down in glue! In that case, top-end effects are largely dissipated before they reach the lower end.

Definition of *very large damping* is somewhat arbitrary. Here, distributed damping is considered to be very large if $\gamma > 0.3$. The maximum resonant displacement is then less than 3.3 times the top-end displacement (see below Equation 14.31).

Table 14–3 gives data and principal results for two cases of a particular riser example. The two left-hand columns of the table give the responses for the riser excited in mode 3, with moderately large damping ($\lambda_D = 0.1$ kN-s/m^2, for which $\gamma = 0.063$). The two right-hand columns give the responses for the riser excited in mode 5, with very large damping ($\lambda_D = 0.5$ kN-s/m^2, for which $\gamma = 0.308$). Note that the test results mentioned in the final section of this chapter give typical linearized values of $\lambda_D < 0.025$ kN-s/m^2, for deepwater drilling risers.

Table 14–3. *Damped vibrations—example data and results*

Riser data	Mass per unit length, m (tonnes/m)		1
	Celerity, c (m/s)		2,500
	Length, L (m)		3,300
	Mass of LMRP, M (tonnes)		200

Case 1		Case 2	
Mode	3	Mode	5
Period T_p (s)	1.12	Period T_p (s)	.62
ω (radians/s)	5.63	ω (radians/s)	10.19
Distributed damping	Damping at equivalent end	Distributed damping	Damping at equivalent end
$\lambda_D = 0.1$ kN-s/m^2	$\gamma = 0.063$	$\lambda_D = 0.5$ kN-s/m^2	$\gamma = 0.308$
$\lambda_B = 0$	L' = 3488m	$\lambda_B = 0$	L' = 3468m

Principal results

	Case 1		Case 2	
U_{max} / U_0	15.93	15.98	3.21	3.25
$T_{max} / mc\omega U_0$	15.98	15.98	3.25	3.25
$T_{top} / U_0\omega^2\Sigma M$	2.03	2.03	.23	.23
U_L / U_0	14.54	14.57	2.40	2.60

The agreement between the tabulated results obtained by the two calculation methods is excellent, even for the mode 5 case with very large damping. Note that the two ratios—U_{max}/U_0 and $T_{max}/mc\omega U_0$—have the same value (= $1/\gamma$). For the equivalent damping system, that result has been explained below equation (14.31). However, that result is not obvious from equations (14.24) and (14.25), for the distributed damping system. Note that the top-end amplification factors $T_{top}/U_0\omega^2\Sigma M$ are less than 1 for the case with very large damping.

Figure 14–4 shows the amplitude ratios for the displacements (U_x/U_0) and the dynamic tensions ($T_x/mc\omega U_0$) for all points along the riser for the two examples of table 14–3. Figures 14–4*a* and 14–4*b* apply to the mode 3 excitation with moderate damping; differences in the responses for the two models are imperceptible. Figures 14–4*c* and 14–4*d* apply to the mode 5 excitation with very large damping; differences between the two models can be clearly seen, and they increase with depth, as expected.

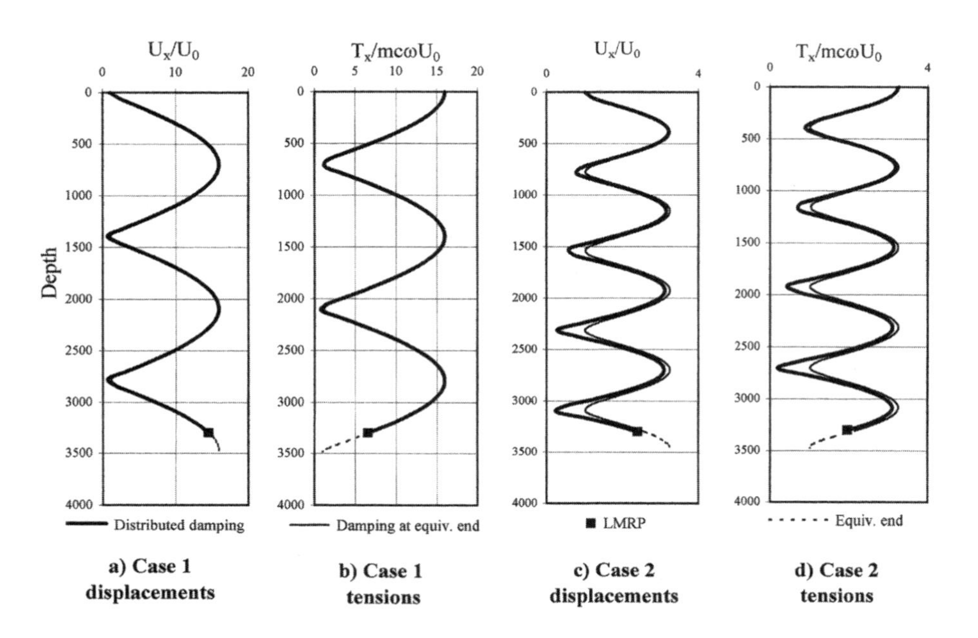

Fig. 14–4. *Riser responses for the examples of table 14–3*

The work sheets of *Hungoff-Damped-Vibrations.xls* can be used to confirm that damping has a significant effect on responses only when the riser is excited close to a resonant frequency. Away from resonance—and particularly for long period excitation—damping can be neglected. The *Forced-Vibrations* work sheet gives the modal proximity *n,* calculated as follows: $n = [(wL'/c)/\pi] + (1/2)$. At resonance, *n* is the mode number. Note that the fluctuations that appear in the responses for very high modes (≥50) are the result of interference with the number of plotted points.

The good agreement between the responses for the two models shows that the riser responses depend essentially on the *quantity* of energy that is dissipated. It is immaterial *how* or *where* that energy is dissipated. This may seem remarkable since the riser may be several kilometers long, and in the equivalent system, the energy is dissipated at an *equivalent point* that may be several hundred meters below the riser bottom end!

Results obtained in this chapter show that accurate responses of a hung-off riser can be obtained using very simple models both for the riser architecture and for the damping system, providing that the latter dissipates the correct quantity of energy. The only remaining problem is to give values to the top-end excitation and the damping at near-resonant frequencies.

Hung-Off Riser Experience and Research Campaigns

Several publications mention the problem of large axial forces induced in hung-off risers in storm conditions. Kwan and colleagues described storm damage to three hung-off risers in 1975.[2] In 1983, a tension fluctuation of 250 tonnesf (2,450 kN) was measured at the top end of a 1,250 m riser hung off in a storm in the Mediterranean Sea.[3] Brekke and colleagues reported a tension fluctuation of 425 kips (1,890 kN) measured at the top end of a 7,700 ft (2,350 m) riser hung off the *Glomar Explorer* in a storm in 1998.[4] Such conditions induce large-amplitude riser top-end displacements (U_0) at periods well in excess of the riser fundamental period. Long-period vessel heave (i.e., long compared to the riser fundamental resonant period) can be obtained sufficiently accurately from sea state data and the vessel's response amplitude operator (RAO). Damping can be neglected. The displacements and the dynamic tensions can then be calculated accurately using the equations of this chapter for the models of figure 14–1*a* through *d*. Such dynamic tensions can be significantly reduced by hanging the riser from the tensioners.[5] That has the effect of greatly reducing the top-end displacement U_0.[6]

The effect of *riser resonance* is more difficult to determine. The natural period of an 1,800 m hung-off drilling riser was measured to be close to 3 s, which corresponded well with the theoretical value.[7] For a 3,000 m drilling riser of comparable architecture, the natural period is on the order of 5–6 s. The resonant response of such a riser depends on the top-end excitation resulting from the drilling-vessel heave at and around that period and the magnitude of the riser damping, neither of which is known precisely.

The Institut Français du Pétrole, with partners, undertook two research campaigns in the 1980s to investigate drilling-vessel short-period heave and riser axial damping.[8] Since then, drilling vessels and deepwater risers have both evolved; thus, results of those campaigns should be used with caution.

Measurements of vessel short-period heave made during the second campaign were more complete. An instrument called a *numerical pallograph* was installed on the Ocean Drilling Program vessel *Joides Resolution* (displacement 16,900 tonnes) from March 1988 to January 1990. During that time, 2,364 sequences of vessel movements, each 40 min, were recorded. Short-period energy, at periods less than 5 s were found to be non-negligible.[9]

Axial damping was investigated during both campaigns. Damping at the LMRP level can be estimated using Morison's equation. Hydrodynamic damping of the main length of riser can be decomposed into two parts: *wall drag,* due to viscous forces acting on the large uniform surfaces of the riser; and *form drag,* resulting from discontinuities in the riser cross-section at the connectors, particularly as a result of discontinuities in buoyancy units.

According to the literature,[10] the wall drag $f_{\mathrm{d\,wall}}$ is approximated by the following formula:

$$f_{\mathrm{d\,wall}} = -\frac{1}{2} C_{\mathrm{d\,wall}} \rho U_x^2 \omega^2 \sin \omega t \qquad (14.34)$$

where the nondimensional coefficient $C_{\mathrm{d\,wall}}$ can be approximated in terms of the Reynolds number $R_{\mathrm{e\,wall}}$; that is,

$$C_{\mathrm{d\,wall}} = \max\left(\sqrt{\frac{2}{R_{\mathrm{e\,wall}}}} ; 0.005 \right) \qquad (14.35)$$

In both campaigns, tests were performed on scale models to determine the form drag at the connectors. Regression analyses of results obtained with a half-scale model during the first campaign gave the following (scaled-up) amplitudes for the drag force F_{d} and the inertia force F_{I}, which contributes to the *added mass*:[11]

$$F_{\mathrm{d}} = 0.30\left(U_x^{1.5}\omega^2\right)\ kN \qquad (14.36)$$

$$F_{\mathrm{I}} = 0.21\left(U_x^{1.33}\omega^2\right)\ kN \qquad (14.37)$$

During the second campaign, tests were performed on a three-quarter–scale model. Regression analyses of the results finally gave the following as the best fit for the form drag at the connectors:[12]

$$F_{\mathrm{d}} = -\frac{1}{2} C_{\mathrm{ld}} \rho (\Sigma S) U_x^2 \omega^2 \sin \omega t \qquad (14.38)$$

where ρ is the fluid density and ΣS is the sum of the surfaces locally opposed to the longitudinal flow; C_{ld} is the longitudinal drag coefficient, which is expressed in equation (14.39) in terms of the Keulegan-Carpenter number K_{c} associated with the longitudinal flow:

$$C_{\mathrm{ld\,wall}} = \frac{3.75}{\sqrt{K_{\mathrm{c}}}} = 3.75\sqrt{\frac{\phi_{\mathrm{H}}}{2\pi U_x}} \qquad (14.39)$$

where ϕ_H is the riser hydraulic diameter associated with form drag (i.e., the difference between the diameters of the buoyancy units and the riser tube).

Risers also experience structural damping, which, for steel, was estimated to be on the order of 2% of critical. Simulations of a 3,000 m riser subject to heave sequences recorded on the *Joides Resolution,* with damping modeled according to equations (14.34) – (14.37), showed that the response was significantly influenced by resonance.[13] The effect of the internal fluid column was not taken into account. If the riser is open at the bottom end, relative movement between the riser and the internal fluid column will be a further cause of wall drag. Dynamic pressures below the LMRP will probably cause the internal fluid column to oscillate in opposition to the riser movement, hence increasing the drag. For a riser with a closed bottom end, pressure (stress) waves will be transmitted by the LMRP to the internal fluid column and vice versa. The fluid column will probably behave as an extension of the riser, folded back within itself—and hence lengthen the resonant periods—although this has not been verified by measurements. There are still unanswered questions about the effect of axial resonance on the longitudinal behavior of hung-off risers.

Summary

This chapter has presented analytical equations for determining the resonant frequencies and the amplitudes of the axial displacements and dynamic tensions of hung-off risers, which may be uniform or composed of segments with different characteristics. It has been shown that the mass of the LMRP causes the riser to respond as though the riser length were increased to an equivalent length L'. It has also been shown that the riser response is governed by the propagation of axial stress waves along the riser. Simple equations have been derived to show that the fundamental resonant period is equal to the time required for a stress wave to run four times the length of the riser.

The accompanying Excel file *Hungoff-Free-Vibrations.xls* allows the propagation of the stress waves to be observed. It gives the riser response for different data, as well as the resonant periods and equivalent lengths L' for different modes.

It has been shown that riser segments can be modeled accurately using mean values for mass, stiffness, and celerity, even if the segments are made up of riser joints composed of *tubes* and *connectors* made from different materials, with quite different characteristics.

Two damping models have been compared using the Excel file *Hungoff-Damped-Vibrations.xls*. In one, the damping is distributed along the riser and also applied at the level of the LMRP at the lower end. In the other, the damping is concentrated at an equivalent point below the riser bottom end. It has been shown that the responses are generally almost identical, providing that the rate of dissipation of energy is the same for both models. It has been confirmed that, away from resonance, damping has negligible influence on riser responses.

The difficult problems of the evaluation of short-period vessel heave and near-resonant damping have been discussed. Published results of research campaigns have been summarized.

References

[1]Kwan, C. T., T. L. Marion, and T. N. Gardner. 1979. Storm disconnect of deepwater drilling risers. Paper presented at the ASME Winter Annual Meeting, New York;
Schawann, J. C., and Sparks C. P. 1985. Riser instrumentation—deep water drilling campaign results. Paper presented at the Deep Offshore Technology Conference, Sorrento, Italy;
Brekke, J. N., B. Chou, G. S. Virk, and H. M. Thompson. 1999. Behavior of a drilling riser hungoff in deep water. Paper presented at the Deep Offshore Technology Conference, Stavanger;
Ambrose, B. D., F. Grealish, and K. Whooley. 2001. Soft hangoff method for drilling risers in ultra deepwater. Paper OTC 13186, presented at the Offshore Technology Conference, Houston;
Sparks, C. P., J. P. Cabillic, and J. C. Schawann. 1982. Longitudinal resonant behavior of very deep water risers. Paper OTC 4317, presented at the Offshore Technology Conference, Houston;
Olagnon, M., O. Besançon, J. Guesnon, F. Mauviel, and C. Charles. 1989. Experimental study of design conditions for very long risers under a floating support. Paper presented at the Deep Offshore Technology Conference, Marbella;

Besançon, O., C. P. Sparks, and J. Guesnon. 1992. Longitudinal resonant behavior of deepwater risers calculated using recorded heave sequences. Paper presented at the 11th Offshore Mechanics and Arctic Engineering Conference, Calgary.

[2]Kwan, Marion, and Gardner, 1979.

[3]Schawann and Sparks, 1985.

[4]Brekke et al., 1999.

[5]Ibid.

[6]Ambrose, Grealish, and Whooley, 2001.

[7]Schawann and Sparks, 1985.

[8]Besançon, Sparks, and Guesnon, 1992.

[9]Ibid.

[10]Ibid.

[11]Sparks, Cabillic, and Schawann, 1982.

[12]Besançon, Sparks, and Guesnon, 1992.

[13]Ibid.

15

Transverse Modal Vibrations of Near-Vertical Risers

In this chapter, the physics of near-vertical riser transverse modal vibrations, of the type known as *vortex-induced vibrations* (VIV), are examined.[1] The simplified analytical solutions presented for such vibrations in this chapter lead to simple expressions for the principal points of concern, including the natural frequencies, the positions of the nodes and antinodes, the maximum angular movements at the riser lower end, and the positions and values of maximum riser curvature. These expressions increase understanding of VIV of deepwater risers under lock-in conditions. The Excel files *Uniform-Transverse-Modal.xls* and *Segmented-Transverse-Modal.xls* enable the reader to compare the results of the different analytical methods described in the text with those obtained by numerical analyses.

Note that again *tension* always implies effective tension.

Physics of Undamped Transverse Vibrations

When an undamped riser is forced to vibrate laterally, transverse waves descend and ascend the riser in a fashion similar to the waves described in chapters 13 and 14 for axial vibrations. In the case of transverse waves, however, the celerity depends on the tension, and hence it is not constant unless the apparent weight is zero. As a result, the nodes are generally not equally spaced, and the vibrations at the antinodes are generally not of equal amplitude.

Figure 15–1 illustrates the case of a riser with positive weight in water and a very low top tension factor, vibrating in the third mode. The descending and ascending waves combine to give the riser lateral displacement at eight equal time intervals during a half modal period. For clarity, one wave crest of each of the two component waves is highlighted with beads. Figure 15–1*a* and *i* show the riser in the mean position; the descending and ascending waves pass each other and cancel completely, with the result that the riser is straight. In this position, the crests of the ascending and descending waves pass each other *at the nodes.*

As the descending and ascending waves progress along the riser, they evolve both in amplitude and wavelength, as a result of the nonconstant tension. Figure 15–1*e* shows the riser in a position of maximum lateral displacement; the two waves combine perfectly and cannot be distinguished from each other. In this position, the crests of the waves pass each other *at the antinodes.*

Despite the nonconstant celerity, the time required for the descending and ascending waves to travel between adjacent nodes (fig. 15–1*a* through *i*) is seen to be equal to half the modal period. Likewise, the time for them to travel between a node and an adjacent antinode (or vice versa) is equal to a quarter of the modal period (fig. 15–1*a* through *e* and *e* through *i*). For the fundamental (mode 1), the nodes are at the extremities. It follows that the period of the fundamental is equal to the time required for the waves to travel twice the length of the riser, just as it was for the axial vibrations of chapter 13.

Vortex-Induced Modal Vibrations

Vortex-induced modal vibrations are slightly different from theoretical modal vibrations because of the particularities of VIV excitation and damping. Experimental data have shown that the vortex shedding frequency for a fixed cylinder is a function of the ratio of current velocity U to diameter D, as given by the Strouhal relation:[2]

$$f_s = S_t \frac{U}{D} \qquad (15.1)$$

where the Strouhal number is generally about 0.2. However, if the cylinder is elastically supported and has a natural frequency close to the one given by equation (15.1), the vortex shedding frequency will adapt itself to

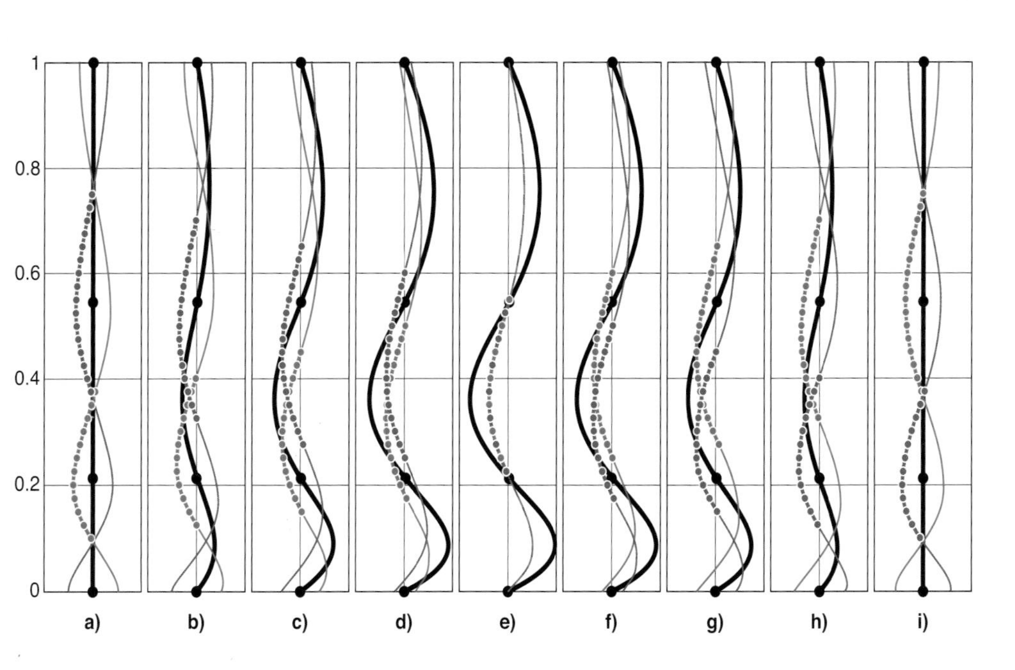

Fig. 15–1. *Riser mode 3 vibration—evolution of descending (blue) and ascending (red) waves, and resultant riser profile (black) during one half modal period*

the natural frequency of the cylinder, and the Strouhal relation will be modified. This situation is known as *lock-in.*

Experimental data has shown that under lock-in conditions, the cylinder vibrates at one of its natural frequencies, but with a (single) amplitude auto-limited to about one diameter. For larger amplitudes, the vortices cease to excite the motion. They have a damping effect.

This limit in amplitude is also considered to apply to risers, although there is little available field data and what there is, is difficult to interpret. Published data from the Norwegian Deepwater Program are particularly significant.[3] They have revealed that the response of four instrumented risers was principally at low modes (1–4). Lock-in was recorded in the second mode of the Helland-Hansen riser, with a modal period of about 20 s.

Hence, when subjected to VIV, riser responses at a natural frequency do not increase indefinitely, and amplitudes of the crests of the lateral movements are all approximately equal (±1 diameter). Therefore, unlike figure 15–1, these amplitudes do not increase with depth. VIV lock-in on a riser causes the mode shape to be modified by a continual addition of energy in the upper sections and a continual dissipation of energy in the lower sections.

Basic Equations for Vibration of a Beam under Constant Tension

The riser can again be considered to be a vertical tensioned beam, for which the basic equation governing transverse vibrations is given by

$$EI\frac{\partial^4 y}{\partial x^4} - T\frac{\partial^2 y}{\partial x^2} - w\frac{\partial y}{\partial x} + m\frac{\partial^2 y}{\partial t^2} = 0 \qquad (15.2)$$

where x and y are the vertical and horizontal axes, EI is the bending stiffness, T is the tension, w is the apparent weight per unit length, and m is the mass (plus added mass) per unit length. Equation (15.2) has an exact analytical solution only for the case of constant tension ($w = 0$). The solution is then given by

$$y = Y_{\rm a} \sin\left(\frac{n\pi x}{L}\right) \sin \omega_n t \qquad (15.3)$$

where $Y_{\rm a}$ is the amplitude of the vibration.

For mode n, the natural frequency ω_n and the corresponding natural period T_{pn} are then given as follows:[4]

$$T_{pn} = \frac{2\pi}{\omega_n} = \frac{2L/n}{\sqrt{(T/m) + (n\pi/L)^2 (EI/m)}} \qquad (15.4)$$

The beam therefore responds exactly as if it were a cable subject to an increased tension T' given by

$$T' = T + \left(\frac{\pi}{L_n}\right)^2 EI \qquad (15.5)$$

where L_n is the internodal length ($L_n = L/n$).

Given that the transmission time of transverse waves between adjacent nodes is equal to half the modal period, as explained in the preceding section, equation (15.4) can be rewritten as

$$T_{pn} = \frac{2L_n}{\sqrt{c_{\text{cable}}^2 + c_{\text{beam}}^2}} \qquad (15.6)$$

where the respective cable and beam celerities are given by equations (15.7) and (15.8):

$$c_{\text{cable}} = \sqrt{\frac{T}{m}} \qquad (15.7)$$

$$c_{\text{beam}} = \frac{\pi}{L_n}\sqrt{\frac{EI}{m}} \qquad (15.8)$$

The combined celerity c in the tensioned beam is then given by $c = \sqrt{c_{\text{cable}}^2 + c_{\text{beam}}^2}$. For a deepwater riser, beam celerity is generally small in comparison to cable celerity and has little influence. For example, for a steel riser of length 2,000 m, diameter 21 in., wall thickness 1 in., and linear mass 1.2 tonnes/m under 1,470 kN constant tension, vibrating in the 10th mode, the combined celerity c is only 2.6% greater than c_{cable}.

For less severe cases, such as risers of smaller diameter, with lower bending stiffness, vibrating at lower modes, with higher axial tensions, the influence of bending stiffness is even less. Consequently, bending stiffness generally has small influence on the natural period of lateral vibrations of a vertical *weightless* deepwater riser, except for very high modes. It is reasonable to expect the same to be true of vertical risers that are not weightless (cf. the influence of EI on the riser static profile in chap. 8). Hence, as a first approximation, the riser will be treated as a cable.

Bessel Cable Analysis of a Riser without Bending Stiffness ($EI = 0$)

For simplicity and clarity, the method of analysis described in this section is called *Bessel cable analysis,* since bending stiffness is neglected and the solution involves Bessel functions. If bending stiffness is neglected, equation (15.2) reduces to

$$T_x \frac{\partial^2 y}{\partial x^2} + w \frac{\partial y}{\partial x} - m \frac{\partial^2 y}{\partial t^2} = 0 \qquad (15.9)$$

Putting

$$z_x = \frac{2\sqrt{mT_x}}{w} \omega_n \qquad (15.10)$$

and

$$y = Y_x \sin \omega_n t \qquad (15.11)$$

where Y_x is the mode shape, allows equation (15.9) to be rewritten as

$$\frac{d^2 Y_x}{dz^2} + \frac{1}{z_x} \frac{dY_x}{dz} + Y_x = 0 \qquad (15.12)$$

which has the following solution:

$$Y_x = AJ_0(z_x) + BY_0(z_x) \qquad (15.13)$$

where $J_0(z_x)$ and $Y_0(z_x)$ are Bessel functions of the first and second kind, of order zero. Constants A and B can be found by iteration, since Y is zero at the extremities.

Although equations (15.11) and (15.13) give the exact solution of equation (15.9), they are not of great practical value, since Bessel functions are difficult to evaluate. Solutions have to be found by iteration. Nevertheless, Bessel cable analysis is interesting and important, since it gives the *exact* solution for a vibrating cable with zero bending stiffness.

Simple Cable Analysis of a Riser without Bending Stiffness ($EI = 0$)

This section describes a method of analysis that is termed *simple cable analysis* for clarity. A very much simpler expression (than equation [15.13]) for the mode shape of a riser with zero bending stiffness can be found by making a slight modification to equation (15.9). This plainly needs to be done with great care if the results are to be of any value. The validity of such a modification can be verified by comparing results with those obtained with the Bessel analysis (equation [15.13]) and with results of numerical analyses. This is done later in this chapter (in the section "Validation Using Excel File *Uniform-Transverse-Modal.xls*").

The three terms on the left-hand side of equation (15.9) represent lateral force components per unit length with different significance, at height x above the riser foot (see fig. 15–2). The first term is the restoring force resulting from the effect of axial tension acting in the curved riser. This is opposed by the inertia force given by the third term. These forces are of constant direction between adjacent nodes.

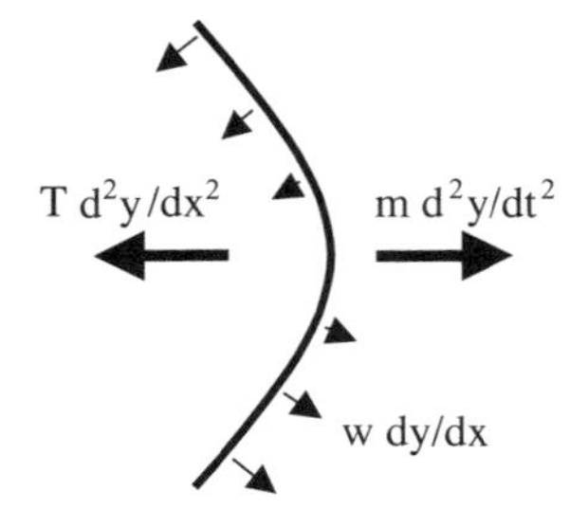

Fig. 15–2. *Lateral force components acting between nodes*

The second term of equation (15.9) is the horizontally resolved component of apparent weight acting perpendicular to the riser axis (see fig. 15-2). Since the riser is near vertical, this term must already be small. Furthermore, this force component changes direction between nodes. Above the antinode, it acts with the first term. Below the antinode, it acts in opposition. When integrated between adjacent nodes, the resultant is zero. Hence, this term must be of much less significance than the other two.

Without further justification, it is proposed to halve the second term of equation (15.9), to obtain an equation with an analytical solution expressed in terms of trigonometric functions instead of Bessel functions. The modified equation is as follows:

$$T_x \frac{\partial^2 y}{\partial x^2} + \left(\frac{1}{2}\right) w \frac{\partial y}{\partial x} - m \frac{\partial^2 y}{\partial t^2} = 0 \qquad (15.14)$$

By use of the substitutions given by equations (15.10) and (15.11), equation (15.14) then leads to

$$\frac{d^2 Y_x}{dz^2} + Y_x = 0 \qquad (15.15)$$

for which the solution given by

$$Y_x = Y_a \sin(z_x - z_b) \qquad (15.16)$$

where Y_a is the modal amplitude (a constant); Y_x is the mode shape, and z_b is the value of z_x (see equation [15.10]) at the riser bottom end.

Note that rather than use physical arguments to justify modifying the second term of equation (15.9), it can alternatively be simply argued that equation (15.12) tends toward equation (15.15) as z_x increases.[5]

The amplitudes at the antinodes according to equation (15.16) are all equal to Y_a. This is in contrast to the Bessel expression of equation (15.13), for which they increase in amplitude as they descend the riser, as seen in figure 15–3.

Since vertical tensioned risers subject to VIV under lock-in conditions do tend to vibrate with virtually equal amplitudes at the antinodes, the simplified solution of equation (15.16) is in fact a closer representation of vortex-induced modal response than the more exact Bessel expression of equation (15.13). Equation (15.16) allows a wealth of important parameters related to lateral vibrations of vertical tensioned risers to be calculated very simply and directly, without iteration.

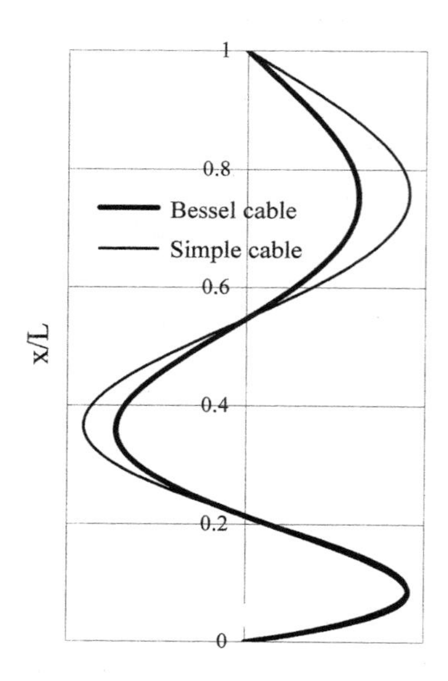

Fig. 15–3. *Example comparison of mode shapes*

Resonant frequencies, periods, and riser mean celerity

For mode n, the natural frequency ω_n and the corresponding period T_{pn} can be found by considering the riser extremities, where the lateral displacement y is zero. For mode n, equation (15.16) gives

$$z_t - z_b = n\pi \tag{15.17}$$

where z_t and z_b are the values of z at the riser top and bottom ends. Substitution of equation (15.10) into equation (15.17) leads to equation (15.18), which gives the natural frequencies ω_n according to the mode number n:

$$\frac{2\sqrt{m}}{w}\left(\sqrt{T_t} - \sqrt{T_b}\right)\omega_n = n\pi \tag{15.18}$$

Since $wL = T_t - T_b$, the corresponding natural periods T_{pn} are given by

$$T_{pn} = \frac{2\pi}{\omega_n} = \frac{4\sqrt{m}}{w}\left(\frac{\sqrt{T_t} - \sqrt{T_b}}{n}\right) = \frac{4L/n}{\sqrt{T_t/m} + \sqrt{T_b/m}} \tag{15.19}$$

Note that coherent units, such as SI units, must be used when using equation (15.19) to evaluate ω_n and T_{pn}.

Since the period T_{p1} of the fundamental (mode 1) is equal to the time for transverse waves to travel twice the length of the riser, the riser mean celerity c_{Rm} is given by

$$c_{Rm} = \frac{2L}{T_{p1}} \tag{15.20}$$

Hence, from equation (15.19),

$$c_{Rm} = \frac{\sqrt{T_t / m} + \sqrt{T_b / m}}{2} \tag{15.21}$$

Thus, the riser mean celerity is the mean of the celerities at the riser extremities.

Finally, the simplest way to determine the natural period of the fundamental (mode 1) is to calculate the mean celerity from equation (15.21) and then determine the period of the fundamental (T_{p1}) from

$$T_{p1_1} = \frac{2L}{c_{Rm}} \tag{15.22}$$

From equation (15.19), the natural period T_{pn} for mode n is then given by

$$T_{pn_n} = \frac{T_{p1_1}}{n} \tag{15.23}$$

Evaluation of parameter z_x

The value of z_x (at height x) can be found from the corresponding tension T_x and the mode number. From equation (15.10), the ratio $z_x / \sqrt{T_x}$ is constant for any particular frequency ω_n. Hence,

$$z_x = z_t \sqrt{\frac{T_x}{T_t}} = z_b \sqrt{\frac{T_x}{T_b}} \tag{15.24}$$

Therefore, $z_x\left(\sqrt{T_t}-\sqrt{T_b}\right) = (z_t - z_b)\sqrt{T_x}$, hence from equation (15.17), for mode n,

$$z_x = \frac{\sqrt{T_x}}{\sqrt{T_\mathrm{t}}-\sqrt{T_\mathrm{b}}} n\pi \qquad (15.25)$$

Positions of nodes and antinodes

At the nodes (subscript "k"), the riser lateral displacement Y is zero. Hence, from equation (15.16), at node k,

$$z_k - z_\mathrm{b} = k\pi \qquad (15.26)$$

where k takes integer values between zero and n, the mode number. For adjacent nodes (subscripts "k" and "$k + 1$"),

$$z_{k+1} - z_k = \pi \qquad (15.27)$$

Thus, from equations (15.25) and (15.26),

$$\sqrt{T_{k+1}} - \sqrt{T_k} = \frac{\sqrt{T_\mathrm{t}}-\sqrt{T_\mathrm{b}}}{n} \qquad (15.28)$$

At the antinodes (subscript "a"), the lateral displacements are maximum. Hence, from equation (15.16),

$$z_a - z_\mathrm{b} = (2a-1)\frac{\pi}{2} \qquad (15.29)$$

where a takes integer values between 1 and n, the mode number. From equations (15.26) and (15.29), for the antinode a between adjacent nodes k and $k + 1$,

$$z_{k+1} - z_a = z_a - z_k = \frac{\pi}{2} \qquad (15.30)$$

Therefore, from equation (15.25),

$$z_x\left(\sqrt{T_t}-\sqrt{T_b}\right) = (z_t - z_b)\sqrt{T_x} \qquad (15.31)$$

Hence, for mode n, the scale of $\sqrt{\text{tension}}$, between the riser top and bottom ends is divided into $2n$ equal intervals by all the nodes and antinodes. This allows the tensions and positions of all the nodes and antinodes to be found directly.

Mean celerities between adjacent nodes

The transmission time $t_{k,k+1}$ of transverse waves between any two adjacent nodes (subscripts "k" and "$k + 1$") is equal to half the modal period T_{pn} (see last paragraph of section: Physics of Undamped Tranverse Vibrations). Hence, from equation (15.19),

$$t_{k,k+1} = \frac{T_{pn}}{2} = \frac{2\sqrt{m}}{w}\left(\frac{\sqrt{T_t} - \sqrt{T_b}}{n}\right) \qquad (15.32)$$

$L_{k,k+1}$ is used to denote the length between the two nodes; therefore $wL_{k,k+1} = T_{k+1} - T_k$; thus, from equation (15.28),

$$t_{k,k+1} = \frac{2\sqrt{m}}{w}\left(\sqrt{T_{k+1}} - \sqrt{T_k}\right) = \frac{2\sqrt{m}}{\sqrt{T_{k+1}} + \sqrt{T_k}} L_{k,k+1} \qquad (15.33)$$

However, if T_a is the tension at the intermediate antinode, then from the first equality of equation (15.31),

$$\sqrt{T_a} = \frac{\sqrt{T_{k+1}} + \sqrt{T_k}}{2} \qquad (15.34)$$

Hence, from equation (15.33),

$$t_{k,k+1} = \frac{L_{k,k+1}}{\sqrt{T_a / m}} \qquad (15.35)$$

Thus, the mean celerity between adjacent nodes is given by

$$c_m = \frac{L_{k,k+1}}{t_{k,k+1}} = \sqrt{\frac{T_a}{m}} = \frac{\sqrt{T_{k+1}/m} + \sqrt{T_k/m}}{2} \qquad (15.36)$$

The mean celerity obtained using equation (15.36) is equal to the celerity at the intermediate antinode, which, from equation (15.34), is equal to the mean of the celerities at the nodes. This result can also be obtained by *assuming* that the celerity at all points is given by $c_x = \sqrt{T_x / m}$ and then integrating between adjacent nodes to find the mean value c_m.

Bottom-end angle

The slope of the riser, in the position of maximum deflection, can be found by differentiating equation (15.16), which gives

$$\frac{dy}{dx} = Y_{a}\sqrt{\frac{m}{T_{x}}}\omega_{n}\cos(z_{x} - z_{b}) \qquad (15.37)$$

Since, at the nodes, $\cos(z_x - z_b) = \pm 1$, the amplitudes of the angles there are given by

$$\left(\frac{dy}{dx}\right)_{nodes} = \pm Y_{a}\sqrt{\frac{m}{T_{x}}}\omega_{n} \qquad (15.38)$$

Therefore, the bottom-end angle is given by

$$\theta_{b} = \pm Y_{a}\sqrt{\frac{m}{T_{b}}}\omega_{n} \qquad (15.39)$$

Note that the riser bottom-end angle as given by equation (15.39) depends on four parameters, none of which is related to the water depth or the riser length. This result is particularly striking for drilling risers. Amplitude Y_a is approximately equal to the riser hydrodynamic diameter (i.e., the diameter of the buoyancy modules). The bottom-end tension of a drilling riser is normally chosen to be just sufficient to lift the LMRP clear, in case of a riser disconnect. Hence, it is independent of water depth. Likewise, the riser linear mass (including contents and added mass) does not increase necessarily with water depth. The natural frequency that may be excited depends on the riser hydraulic diameter and current velocity according to the Strouhal relation (see equation [15.1]).

Equation (15.39) can be further simplified. From equation (15.32),

$$\omega_{n} = \frac{2\pi}{T_{pn}} = \frac{\pi}{t_{k,k+1}} \qquad (15.40)$$

$L_{b,b+1}$ is used to denote the length between the riser bottom end and the first node above it, and T_{a1} is used to denote the tension at the intermediate antinode; thus, from equations (15.35) and (15.40),

$$\omega_{n} = \pi\frac{\sqrt{T_{a1}/m}}{L_{b,b+1}} \qquad (15.41)$$

Substituting into equation (15.39) gives

$$\theta_b = \pm Y_a \frac{\pi}{L_{b,b+1}} \sqrt{\frac{T_{a1}}{T_b}} \tag{15.42}$$

Finally, the bottom-end angle depends on geometrical parameters of the riser profile (riser amplitude and height of first node) amplified by the factor $\sqrt{T_{a1}/T_b}$.

Riser curvature (1/R) at the antinodes

Since riser curvature $1/R = d^2y/dx^2$, differentiation of equation (15.37) yields

$$\frac{1}{R} = Y_a \frac{m\omega_n^2}{T_x}\left[\sin(z_x - z_b) + \frac{\cos(z_x - z_b)}{z_x}\right] \tag{15.43}$$

At the antinodes, the bracketed term on the right-hand side of equation (15.43) is equal to 1; thus, the curvature is given by

$$\frac{1}{R} = Y_a \frac{m\omega_n^2}{T_a} \tag{15.44}$$

Since tension decreases with depth, equation (15.44) has its greatest value at the lowest antinode, where, from equation (15.41),

$$\frac{1}{R} = Y_a \left(\frac{\pi}{L_{b,b+1}}\right)^2 \tag{15.45}$$

Maximum riser curvature (1/R)

Since riser tension decreases with depth, the maximum curvature will occur for $d^3y/dx^3 = 0$ at some point below the lowest antinode. Differentiation of equation (15.43) shows that this occurs for $z_x = z_{c\,max}$, as given by

$$\tan(z_{c\,max} - z_b) = \frac{z_{c\,max}}{3} - \frac{1}{z_{c\,max}} \tag{15.46}$$

The value of $z_{c\,max}$ depends only on z_b and can be found by iteration. The tension corresponding to $z_{c\,max}$ can then be found from equation

(15.25). Note that since z_b increases with the mode number (see equation [15.10]), as it does so, $z_{c\,max}$ tends increasingly toward $z_b + (\pi/2)$, which is the value of z at the first antinode.

Simple Beam Analysis of a Riser with Bending Stiffness ($EI \neq 0$)

This section describes an analysis method that is called *simple beam analysis* for clarity. In the section on the vibration of beams under *constant tension,* it was shown that the beam behaved precisely as a cable subject to a tension increased by $(\pi/L_n)^2EI$, where L_n is the internodal length (see equation [15.5]). The object now is to see if that approach can also be used, at least approximately, for a beam under uniformly *varying tension.* Equation (15.2) with the apparent weight term on the left-hand side again halved, as in equation (15.14), becomes

$$EI\frac{d^4y}{dx^4} - T_x\frac{d^2y}{dx^2} - \left(\frac{1}{2}\right)w\frac{dy}{dx} + m\frac{d^2y}{dt^2} = 0 \qquad (15.47)$$

The procedure consists of first using a modified value of tension (T'_x) in equation (15.10), as given by

$$T'_x = T_x + Q \qquad (15.48)$$

where Q is a constant to be determined, for each pair of nodes. For the case of constant tension, $Q = (n\pi/L)^2EI$, and equation (15.48) corresponds to equation (15.5). It is then *assumed* that there is a solution between nodes, of the form

$$y = Y_a \sin(z'_x - z'_k)\sin\omega'_n t \qquad (15.49)$$

Primed values of all parameters (T'_k, T'_a, z', ω'_n, T'_{pn}, c', L'_n, etc.) apply to the case with bending stiffness. Since the assumed form of the solution is the same as for the simple cable analysis, the equations of that section can be applied between pairs of nodes, providing that primed values are used throughout. Substitution of equation (15.49) into equation (15.47) leads to the following requirement:

$$EI\frac{d^2y}{dx^2} + Qy = 0 \qquad (15.50)$$

A mean constant value of Q is required for each length between nodes. This can be found by integrating equation (15.50) between adjacent nodes (k to $k+1$). From equation (15.49), this leads to

$$EI\frac{2m\omega_n'^2}{w}\left[\frac{\cos(z_x'-z_k')}{z_x'}\right]_k^{k+1}=Q\frac{w}{2m\omega_n'^2}\left[z_x'\cos(z_x'-z_k')-\sin(z_x'-z_k')\right]_k^{k+1} \quad (15.51)$$

Hence, from equation (15.27),

$$EI\frac{2m\omega_n'^2}{w}\left(\frac{1}{z_{k+1}'}+\frac{1}{z_k'}\right)=Q\frac{w}{2m\omega_n'^2}\left(z_{k+1}'+z_k'\right) \quad (15.52)$$

Since the transmission time between adjacent nodes is equal to half the modal period, it follows that

$$\omega_n'=\frac{2\pi}{T_{pn}'}=\frac{c_{\mathrm{m}}'}{L_{k,k+1}'}\pi \quad (15.53)$$

Equations (15.10), (15.52), and (15.53) then give

$$Q=\left[\frac{2m\omega_n'^2}{w}\right]^2\frac{EI}{z_k'z_{k+1}'}=m\left(\frac{\pi c_m'}{L_{k,k+1}'}\right)^2\frac{EI}{\sqrt{T_k'T_{k+1}}} \quad (15.54)$$

Finally, from equation (15.36),

$$Q=\frac{T_a'}{\sqrt{T_k'T_{k+1}'}}\left(\frac{\pi}{L_{k,k+1}'}\right)^2 EI \quad (15.55)$$

The factor $T_a'/\sqrt{T_k'T_{k+1}'}$ is equal to unity for the case of constant or zero tension. In fact, it can never differ greatly from 1 for practical riser cases. For the example data of table 15–2, it varies from 1.4, for mode 1 (for which EI has negligible effect), to 1.03, for the length above the riser bottom end for mode 5. This factor approaches 1 ever more closely as the mode number increases.

The mean celerity between adjacent nodes is given by equation (15.36), using the primed values c_{m}' and T_a'. Between nodes k and $k+1$,

$$c_{\mathrm{m}}'=\sqrt{\frac{T_a'}{m}}=\frac{\sqrt{T_{k+1}+Q}+\sqrt{T_k+Q}}{2\sqrt{m}} \quad (15.57)$$

Thus,

$$c'_{\mathrm{m}} = \frac{\sqrt{\left(c_{k+1}\right)^2_{\text{cable}} + c^2_{\text{beam}}} + \sqrt{\left(c_k\right)^2_{\text{cable}} + c^2_{\text{beam}}}}{2} \tag{15.58}$$

where $c_{\text{beam}} = \sqrt{Q/m}$. Tension is lowest between the riser bottom end and the first node. Consequently, that is where cable celerities have their lowest values. It is also where the internodal distance is shortest—and, hence, where Q is greatest (see equation [15.55]) and beam celerity is greatest. Therefore, it is between the riser bottom end and the first node that EI has its greatest effect.

Since the value of Q is different for each pair of nodes, the procedures used for simplified cable analysis cannot be used to find the natural periods and the positions of nodes. Another approach must be used.

The modal period and the distance between nodes can be expressed in terms of the mean celerities between nodes—and, hence, in terms of $\sqrt{T'_a}$ (see equation [15.36]). Since the transmission time between adjacent nodes is equal to half the modal period and since the sum of the internodal lengths $\sum L'_{\mathrm{i}}$ is equal to the riser total length L, it follows that

$$L = \sum L'_{\mathrm{i}} = \sum\left(\frac{T'_{pn}}{2} c'_{m_{\mathrm{i}}}\right) = \frac{T'_{pn}}{2} \sum c'_{m_{\mathrm{i}}} \tag{15.59}$$

Hence, from equations 15.57 and 15.59, the modal period T'_{pn} is given by

$$T'_{pn} = \frac{2L}{\sum c'_{\mathrm{m_i}}} = \frac{2L\sqrt{m}}{\sum \sqrt{T'_a}} \tag{15.60}$$

where $\sum c'_{\mathrm{m_i}}$ is the sum of the mean celerities $c'_{\mathrm{m_i}}$ of all the internodal lengths. Likewise $\sum \sqrt{T'_a}$ is the sum of the values of $\sqrt{T'_a}$ for all the anti-nodes.

$$L'_{k,k+1} = \frac{T'_{pn}}{2} c'_{\mathrm{m}k,k+1} = \frac{c'_{\mathrm{m}k,k+1}}{\sum c'_{\mathrm{m_i}}} L \tag{15.61}$$

Hence, from equations (15.57) and (15.60),

$$L'_{k,k+1} = \frac{\sqrt{T'_a}}{\sum \sqrt{T'_a}} L \tag{15.62}$$

Equation (15.60) gives the modified modal periods. The modified lengths between nodes can be calculated from either equation (15.61) or equation (15.62). The bottom-end angle and the riser curvature at the lowest antinode can then be found from equations (15.42) and (15.45) respectively, using the primed parameters.

The simple beam method of analysis takes into account the bending stiffness with reasonable accuracy, as will be demonstrated in the section "Validation Using Excel File *Uniform-Transverse-Modal.xls*". However, it is complicated, since results have to be obtained by iteration.

Approximate Beam Analysis of a Riser with Bending Stiffness ($EI \neq 0$)

Approximate results can be found very much more easily by using a constant value Q_{mean} for all spans between nodes, instead of solving equation (15.55) for each internodal length. This is called *approximate beam analysis* for clarity. Results are less accurate than with the preceding method, but they are obtained more simply since no iterations are required. The best mean value Q_{mean} is obtained by putting the factor $T'_a / \sqrt{T'_k T'_{k+1}}$ equal to unity and the mean internodal length equal to L/n, where L is the total riser length and n is the mode number. That gives the same value of Q as if tension were constant:

$$Q_{\text{mean}} = \left(\frac{\pi}{L/n}\right)^2 EI \qquad (15.63)$$

A modified value of tension (T') is first defined by adding Q_{mean} to the tension at all points along the riser, according to equation (15.48). Then, with the exception of equation (15.23), equations (15.17)–(15.46) can be used to obtain directly approximate (primed) values of all significant parameters ($T'_k, T'_a, z', \omega'_n, T'_{pn}, c', L'_n$, etc.).

Equation (15.23) cannot be used to determine modal periods T'_{pn}, since Q_{mean} is different for each mode. Instead, the modal periods can be obtained from equation (15.64), where c'_{Rm} is given by equation (15.21), using the primed values for the top- and bottom-end tensions.

$$T'_{pn} = \frac{2L}{n(c'_{\text{Rm}})} \qquad (15.64)$$

Again, the scale of $\sqrt{T'}$ is divided into equal intervals by all nodes and antinodes, which allows their positions to be determined directly. Likewise, the bottom-end angle and the curvature at the lowest antinode can be found directly from equations (15.39) and (15.44), using the primed values ω'_n, T'_b, and T'_a.

Validation Using Excel File *Uniform-Transverse-Modal.xls*

The accuracy of the preceding methods of analysis can be assessed using the Excel file *Uniform-Transverse-Modal.xls,* which calculates results for uniform risers for different data. Results are compared with those obtained by *numerical beam analysis*, in which the riser is modeled with its bending stiffness, by 600 elements, using the finite difference method. Thus, the file *Uniform-Transverse-Modal.xls* allows the results of five different calculation methods to be compared. Table 15–1 summarizes the five methods, along with the approximations on which they rely.

Table 15–1. *Summary of analysis methods and their approximations*

Analysis method	Calculation	Approximations	Basic equation
Bessel cable	Iterative	Bending stiffness neglected (EI=0)	Equation 15.9
Simple cable	Direct	Bending stiffness neglected (EI=0), apparent weight term halved	Equation 15.14
Approximate beam	Direct	Apparent weight term halved, EI treated as mean tension increase, Q_{mean}	Equation 15.47
Simple beam	Iterative	Apparent weight term halved, EI treated as tension increase, Q between nodes	Equation 15.47
Numerical beam	Iterative	No approximations	Equation 15.2

In the names of the methods, *cable* implies that bending stiffness is neglected (i.e., $EI = 0$), and *beam* implies that it is included in the analysis (i.e., $EI \neq 0$). The simple cable and approximate beam methods have the advantage of providing results directly, without iteration.

The principal object of *Uniform-Transverse-Modal.xls* is to demonstrate the accuracy of the simple cable calculation method, since all the important conclusions about resonant periods, node positions, values and positions of maximum curvature, and bottom-end angles are based on equations derived using that method. Comparison of the results of the different methods makes evident the influence of the different approximations.

The *Mode-Shapes* work sheet shows graphically the comparative mode shapes for the displacements and curvature for the five methods. The displacement curves are similar to figure 15–3, although five plots are superimposed, instead of two. To avoid confusion, results of particular calculation methods can be suppressed or have their sign reversed. The graphs are all traced to the same scale, which is not shown. The maximum amplitude of the displacements is defined by the user and is used to determine the maximum curvature and the curvature at the first antinode, as well as the bottom-end angle. The position of the maximum curvature as compared to the position of the first antinode can be observed by comparing the displacement and curvature plots. The *Mode-Shapes* work sheet also gives the natural frequency and the corresponding period for the specified mode (maximum: mode 50), the heights of the lowest node and antinode, the maximum curvature and the curvature at the lowest antinode, and the angle at the riser bottom end.

The *Node-Heights* work sheet allows the heights of all the nodes and antinodes to be compared.

Much more can be deduced by using *Uniform-Transverse-Modal.xls* than can be presented in tables. Nevertheless some example results are included here. Table 15–2 gives data for an example used in previous publications.[6] It corresponds to a drilling riser with 22 in. outer diameter and 1 in. wall thickness, subject to 700 tonnesf top tension, with a top-tension factor of 1.1.

Table 15–2. *Example riser data*

Riser length, L	2,000 m
Bending stiffness, EI	318,600 kN-m^2
Top tension, T_t	7,553.7 kN
Total apparent weight, wL	6,867 kN
Mass + added mass, m	1.2 tonnes/m
Amplitude of vibrations, Y_a	1 m

Table 15–3 compares the results for modal periods between 1 and 50. Results of all five methods compare well up to mode 10. The influence of bending stiffness becomes significant only above about mode 30.

Table 15–3. *Comparative modal periods*

Mode	Bessel cable	Simple cable	Approximate beam	Simple beam	Numerical beam
1	78.79	77.47	77.46	77.45	78.71
2	38.92	38.74	38.71	38.69	38.84
3	25.88	25.82	25.78	25.75	25.78
4	19.39	19.37	19.32	19.26	19.27
5	15.51	15.49	15.43	15.36	15.37
10	-	7.75	7.62	7.52	7.52
20	-	3.87	3.64	3.54	3.55
30	-	2.58	2.28	2.21	2.21
40	-	1.94	1.59	1.54	1.55
50	-	1.55	1.18	1.15	1.15

Table 15–4 gives the node and antinode heights for fifth-mode vibrations. Comparing the results of simple beam and numerical beam analyses reveals that the former method gives the node heights more accurately than the antinode heights. This is because the $w\,dy\,/\,dx$ term in the basic equations (see equations [15.9], [15.14], and [15.47]) *does* contribute to the celerity of the transverse waves. Between any node and the antinode below it, this term combines with the restoring force (see fig. 15–2) and has the effect of slightly increasing celerity. Between the antinode and the node below it, the opposite is the case. Since this term has been halved for simple beam analysis (see equation [15.47]), the effect is reduced, and as a result, the antinodes are raised slightly.

Between two adjacent nodes, the two effects virtually cancel—hence, the good agreement for the positions of the nodes. The results obtained for the nodes by use of the approximate beam method are slightly better than those obtained using the simple cable method but are not as good as those obtained using the simple beam method, as expected.

Table 15–4. *Comparative node (and antinode) heights for fifth-mode vibrations*

Nodes	Bessel cable	Simple cable	Approximate beam	Simple beam	Numerical beam
5	2,000	2,000	2,000	2,000	2,000
	(1,699.3)	(1,703.4)	(1,704.2)	(1,705.6)	(1,701.4)
4	1,428.2	1,428.3	1,429.7	1,432.3	1,432.4
	(1,170.1)	(1,174.6)	(1,176.5)	(1,180.0)	(1,176.0)
3	942.0	942.4	944.6	948.9	949.1
	(727.3)	(731.7)	(734.0)	(738.7)	(734.9)
2	542.1	542.4	544.6	549.5	549.9
	(370.0)	(374.6)	(376.5)	(381.1)	(377.8)
1	228.1	228.3	229.7	233.8	234.4
	(99.4)	(103.4)	(104.2)	(106.3)	(104.5)
0	0	0	0	0	0

Table 15–5 gives the bottom-end angles for different modes 2–20. The influence of bending stiffness can be seen to begin to be significant for mode 10 and above.

Table 15–5. *Comparative bottom-end angles for modes 2–20*

Mode	Bessel cable	Simple cable	Approximate beam	Simple beam	Numerical beam
2	0.49°	0.39°	0.39°	0.39°	0.44°
5	1.08°	0.97°	0.96°	0.94°	0.94°
10	-	1.94°	1.87°	1.72°	1.67°
20	-	3.88°	3.42°	2.92°	2.84°

Table 15–6 gives the maximum curvature for different modes between 1 and 20. The simplified methods give poor results for mode 1, which is of little importance since the curvature is negligible for that mode. The influence of bending stiffness is again significant for mode 10 and above.

Table 15–6. *Comparative maximum curvature for modes 1–20*

Mode	Bessel cable	Simple cable	Approximate beam	Simple beam	Numerical beam
1	0.000003	0.000002	0.000002	0.000002	0.000003
2	0.000019	0.000018	0.000018	0.000018	0.000019
5	0.000190	0.000189	0.000187	0.000181	0.000185
10		0.000923	0.000869	0.000751	0.000748
20		0.004108	0.003268	0.002412	0.002378

Table 15–7 summarizes the simple beam calculation method for the fifth mode, with reference to the relevant equations of this chapter. Note that L_n' is the internodal length. The calculation appears complicated since the values of Q depend on the internodal lengths L_n', which depend on the values of Q. Therefore, an iterative calculation is required. However, convergence is generally rapid, particularly for low-mode numbers.

Table 15–7. *Example* simple beam *calculation for the fifth mode*

	Final values after iteration				Node and (anti-node) heights (m)
	Q	√(T_a')	T_a'	L_n'	
Node number	Eq (15.55)	Equation (15.34) (primed)	From √(T_a')	Equation (15.62)	[nodes from L_n'] [anti-nodes from T_a']
5					2,000.0
	9.81	80.9	6,552.7	567.7	(1,705.6)
4					1,432.3
	13.56	68.9	4,751.9	483.4	(1,180.0)
3					948.9
	19.94	56.9	3,242.8	399.3	(738.7)
2					549.5
	32.11	45.0	2027.3	315.8	(381.1)
1					233.8
	59.47	33.3	1,111.3	233.8	(106.3)
0					0
	Σ =	285.2	Σ =	2,000.0	

Modal period $T_p = 2L\sqrt{m}/\Sigma\sqrt{Ta'} = 2\times2{,}000\times\sqrt{(1.2)}/285.2 = 15.4$ s (Eq.15.60)

Table 15–8 summarizes the complete approximate beam calculation for the fifth mode, with reference to the relevant equations. The approximate beam calculation is very simple since the same value of Q_{mean} is used for all internodal lengths. Calculations for the simple cable method are similar but even simpler, since $Q_{mean} = 0$.

Table 15–8. *Example approximate beam calculation for the fifth mode*

Node number	Q_{mean} Equation (15.63)	$\sqrt{(T')}$ [Equal intervals]	T' From $\sqrt{(T')}$	Tension T' - Q_{mean}	Node and (anti-node) Heights (m) [from tension]
5		87.0	7,573.4	7,553.7	2,000.0
	19.65	(81.0)	(6,557.8)	(6,538.1)	(1,704.2)
4		74.9	5,615.3	5,595.7	1,429.7
	19.65	(68.9)	(4,745.9)	(4,726.3)	(1,176.5)
3		62.8	3,949.6	3,930.0	944.6
	19.65	(56.8)	(3,226.4)	(3,206.7)	(734.0)
2		50.8	2,576.2	2,556.6	544.6
	19.65	(44.7)	(1,999.1)	(1,979.5)	(376.5)
1		38.7	1,495.1	1,475.5	229.7
	19.65	(32.6)	(1,064.2)	(1,044.5)	(104.2)
0		26.6	706.4	686.7	0
Mean celerity $c'_{Rm} = (\sqrt{T_t'} + \sqrt{T_b'})/(2\sqrt{m}) = (87.0 + 26.6)/(2\sqrt{1.2}) = 51.85$ m/s (Eq.15.21)					
Modal period $T'_{pn} = 2L/(n.c'_{Rm}) = 2\times 2{,}000/(5\times 51.85) = 15.4$ s (Eq.15.64)					
Frequency $\omega'_n = 2\pi/T'_{pn} = 2\pi/15.4 = 0.407$ radian/s					
Curvature at 1st anti-node $= Y_a m\omega'^2_n/T'_a = 1\times 1.2\times 0.407^2/1064.2 = 0.00019\ m^{-1}$ (Eq.15.44)					
Bottom end angle $= Y_a\omega'_n\sqrt{(m/T'_b)} = 1\times 0.407\times\sqrt{(1.2/706.2)} = 0.017$ rad $= 0.96°$ (Eq. 15.39)					

Segmented Risers—Modal Responses

The *simple cable analysis* method can be adapted to a segmented riser without difficulty, using an approach similar to that described in chapters 13 and 14. At the junctions between segments, the horizontal displacements and forces must be conserved. The horizontal displacement of each segment can be expressed by equation (15.16) in terms of an *equivalent bottom end,* characterized by parameter z_b''. (Double primes are used to avoid confusion with the single primes already used in this chapter.)

For the horizontal displacement at the bottom end of segment $s + 1$ to be equal to that at the top end of segment s, equation (15.65) must be respected, where the segments are numbered from the lower end:

$$\left[Y_a \sin\left(z_b - z_b''\right)\right]_{s+1} = \left[Y_a \sin\left(z_t - z_b''\right)\right]_s \qquad (15.65)$$

The horizontal force is equal to $T\, dy/dx$, where $dy/\mathrm{d}x$ is given by equation (15.37). For $T\, dy/dx$ at the bottom end of segment $s + 1$ to be equal to $T\, dy/dx$ at the top end of segment s, equation (15.66) must be respected:

$$\left[Y_a \sqrt{m} \cos\left(z_b - z_b''\right)\right]_{s+1} = \left[Y_a \sqrt{m} \cos\left(z_t - z_b''\right)\right]_s \qquad (15.66)$$

Note that tension T is eliminated, since it is the same above and below the junction.

Dividing equation (15.65) by equation (15.66) yields

$$\left[\frac{\tan\left(z_b - z_b''\right)}{\sqrt{m}}\right]_{s+1} = \left[\frac{\tan\left(z_t - z_b''\right)}{\sqrt{m}}\right]_s \qquad (15.67)$$

Equation (16.67) allows all the values of z_b'' to be determined by working upward from the bottom end. For the lowest segment, $z_b'' = z_b$. Once the values of z_b'' have been determined, equation (15.65) can be used to find the amplitude Y_a for each segment. Resonance will occur when, for the top-end segment, $(z_t - z_b'')_{top} = n\pi$, which can be found by iteration. Note that when using equation (15.67) to find values of z_b'' that lead to resonance, it is again important to conserve the same number of integral multiples of π for $(z_b - z_b'')_{s+1}$ as for $(z_t - z_b'')_s$, just as it was in chapters 13 and 14 for axial vibrations of segmented risers.

The transverse displacements (for points within segment s) are then given by equation (15.68) (cf. equation [15.16]):

$$y = \left[Y_a \sin(z_x - z_b'')\right]_s \sin\omega_n t \qquad (15.68)$$

where parameters within the brackets refer to segment s. This can be split into the sum of ascending and descending displacement waves:

$$y = \left[\frac{Y_a}{2}\cos(z_x - z_b'' - \omega_n t) - \frac{Y_a}{2}\cos(z_x - z_b'' + \omega_n t)\right]_s \qquad (15.69)$$

Likewise the curvature (for points within segment s) is then given by equation (15.70) (cf. equation [15.43])

$$\frac{1}{R} = \left\{Y_a \frac{m\omega_n^2}{T_x}\left[\sin(z_x - z_b) + \frac{\cos(z_x - z_b)}{z_x}\right]\right\}_s \sin\omega_n t \quad (15.70)$$

This can also be split into the sum of ascending and descending curvature waves:

$$\frac{1}{R} = \left\{\begin{array}{l} Y_a \dfrac{m\omega_n^2}{2T_x}\left[\cos(z_x - z_b - \omega_n t) - \dfrac{\sin(z_x - z_b - \omega_n t)}{z_x}\right] \\ + Y_a \dfrac{m\omega_n^2}{2T_x}\left[-\cos(z_x - z_b + \omega_n t) + \dfrac{\sin(z_x - z_b + \omega_n t)}{z_x}\right]\end{array}\right\}_s \qquad (15.71)$$

where the upper expression is the ascending wave, and the lower one the descending wave.

Validation Using Excel File *Segmented-Transverse-Modal.xls*

Equations (15.65) and (15.67) have been programmed into the Excel file *Segmented-Transverse-Modal.xls* to model a six-segment riser by using simple cable analysis. The numerical method has also been adapted to the analysis of a six segment riser by using 100 elements to model each of the segments. It gives the *exact* solution, including the effects of bending stiffness.

The *Mode-Shapes* work sheet of *Segmented-Transverse-Modal.xls* displays the modal transverse displacements graphically for both methods, to the same scale. The maximum displacement is entered by the user but may in some cases occur at different antinodes for the two models. Curvatures are also displayed graphically, to the same scale, and their maximum values are given in the results. The positions of the maxima can be noted on the graphs.

The two methods are completely independent, but both are iterative. They are robust and generally converge even when the data include large abrupt changes in characteristics (w, EI, and m), which can cause convergence problems for certain commercial programs. Abrupt changes in characteristics, as in the following example, can produce surprising mode shapes, but with good agreement between the two models. For cases with zero apparent weight ($w = 0$) and zero bending stiffness ($EI = 0$), the results should agree virtually exactly.

Abrupt changes in characteristics between segments can cause discontinuities in the riser curvature at the segment junctions.

Table 15–9. *Example data and principal results for a segmented riser*

Top tension (kN)	**4,400**			
Max. amplitude (m)	**1**	**Apparent weight per unit length**	**Bending stiffness**	**Mass per unit length**
Data per segment	**Length L (m)**	**w (kN/m)**	**EI ($kN\text{-}m^2$)**	**m (tonnes/m)**
Segment 6 (top)	300	0	318,600	5
Segment 5	500	0	318,600	5
Segment 4	400	3	318,600	1
Segment 3	100	3	318,600	1
Segment 2	500	3	318,600	2
Segment 1 (btm)	200	3	318,600	1
Mode 5 results		**Simple cable**	**Numerical beam**	
ω_n (radians/s)		0.2807	0.2813	
T_{pn} (s)		22.38	22.34	
Max. curvature (m^{-1})		0.000078	0.000081	

Table 15–9 gives example data and principal results. These data have been used to generate figure 15–4, which shows the mode 5 displacements and curvature.

The curvature plots are traced twice—with similar signs in the middle sketch and with opposite signs in the right-hand sketch. The former allows the maximum values of displacements and curvature, given by the two methods, to be compared. The latter allows the good agreement between the positions of the nodes to be observed. The discontinuities in curvature at junctions with large abrupt changes in mass per unit length can be clearly seen.

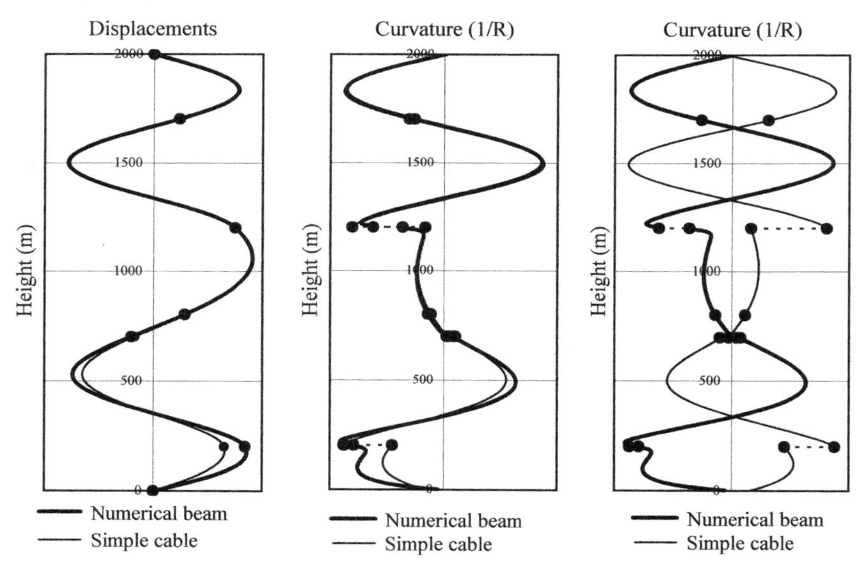

Fig. 15–4. *Example modal displacements and curvature*

The *Waves* work sheet of *Segmented-Transverse-Modal.xls* allows the reader to observe how, for the solution obtained using the simple cable method, the ascending and descending transverse waves propagate and combine to give the riser transverse displacements and curvature (see equations [15.69] and [15.71]). The work sheet allows the reader to advance the time in steps equal to $^1/_{32}$ of the modal period. There can, of course, be no discontinuity in the transverse displacements, but for certain data, discontinuities occur in the *waves* at the segment junctions as a result of partial reflections at the junctions.

At time steps $N = 0$ and 16, the two waves are mirror images of each other. The crests pass each other with opposite signs. The sum is then zero, and the riser is straight. At $N = 8$ and 24, the crests pass each other with the same sign. The two waves combine perfectly and give the maximum displacement, which corresponds to the mode shape.

Extension to Catenary Risers

The preceding sections have all been devoted to near-vertical risers. Some of the same reasoning can nevertheless be applied to catenary risers to obtain the resonant frequencies of transverse vibrations (as opposed to in-plane lateral vibrations). Earlier in this chapter, the fundamental resonant period was seen to be equal to the time required for a transverse wave to descend and reascend the riser. This is expressed by equation (15.22) as being equal to twice the length of the riser, divided by the mean celerity. For the simple cable analysis, the celerity at all points along the riser was equal to $\sqrt{T/m}$ (see following equation [15.36]).

Integration over the length of the near-vertical uniform riser, for which tension varies linearly with depth, gave the exact mean value of the celerity as expressed by equation (15.21). For a catenary riser, the mean celerity can be taken as

$$c_{\mathrm{Rm}} = \frac{\sqrt{T_{\mathrm{t}}} + \sqrt{H}}{2\sqrt{m}} \qquad (15.72)$$

However, equation (15.72) is less exact, since the tension of a catenary does not vary exactly linearly along the suspended length. Nevertheless, equation (15.72) has the merit of simplicity and gives a good approximation of the mean celerity, even for a catenary. If it is assumed that the catenary is pinned at the extremities, equation (15.73) can be used to obtain a quick estimate of the transverse modal periods:

$$T_{\mathrm{p}n} = \frac{2s}{n(c_{\mathrm{Rm}})} \qquad (15.73)$$

where s is the catenary suspended length.

Croutelle and colleagues have given the transverse modal periods for the first five modes of a particular example of a 12¾ in. SCR with a suspended length of 1,931 m.[7] The article gives details of pipe section and fluid densities that allow the catenary weight per unit length to be determined, as well as the "mass plus added mass," assuming an added-mass coefficient $C_m = 1$. The published data for that example are summarized in table 15–10, in which the resulting transverse modal periods are compared with the results deduced from equations (15.72) and (15.73). The resonant periods compared well.

The significance of equations (15.72) and (15.73) goes beyond their accuracy. They allow the influence that platform offset and current exert on transverse modal periods to be understood.

Table 15–10. *Catenary riser example data and modal periods*

	Example data[6]		Units
d	Depth	1,500	m
s	Suspended length	1,931	m
x	Horizontal projection	1,023	m
OD	Outer diameter	12.75	in
w	App.wt. per unit length	0.97	kN/m
m	Mass+added mass	0.26	Tonnes/m
T_t	Top tension	1,933	kN
H	Horizontal tension	478	kN
c_{Rm}	Approx. mean celerity $c_{Rm} = (\sqrt{T_t}+\sqrt{H})/(2\sqrt{m})$	64.55	m/s
	Modal Periods T_p (s)		
Mode 'n'	**Direction**	**Reference[6]**	**$2s/(n.c_{Rm})$**
1	Transverse	59.2	59.8
2	Transverse	29.8	29.9
3	Transverse	20.0	19.9
4	Transverse	15.0	15.0
5	Transverse	12.0	12.0

If platform offset is increased, the tensions (T_t and H) will increase. The mean celerity will therefore increase, which will tend to reduce the modal period. However, the TDP will shift, causing an increase in the suspended length s, which will tend to increase the modal period. These two effects oppose each other; hence, a change in platform offset has little effect on the modal period. Equations (15.72) and (15.73) are also used in the Excel file *TDP-Shift.xls,* referred to in chapter 12, to give approximate values of the transverse modal periods for the platform in the specified near, mean, and far positions, for the mode number input by the user.

In-plane current will cause a slight increase (or decrease, according to the direction) in tension and, therefore, in mean celerity. Hence, it will cause a corresponding slight change in modal period.

No simple method can be given for the calculation of modal periods for *in-plane lateral vibrations,* because of coupling with axial vibrations, particularly at low modes. This can be easily understood by considering first-mode in-plane vibrations. If the catenary ends are fixed, there can be no lateral in-plane vibration without change in axial length. Hence, there is inevitable coupling with axial vibrations. The effect of coupling reduces as the mode number increases.

Summary

This chapter has used different methods to analyze riser transverse modal vibrations of the VIV type under lock-in conditions, for which the amplitude is auto-limiting to approximately one (hydraulic) diameter. The vibrational behavior of beams under constant tension has been recalled, and it has been noted that it is exactly the same as the behavior of a tensioned cable subject to a tension increased by $(\pi/L_n)^2EI$, where L_n is the internodal length, which depends on the mode number (see equation [15.5]).

Modal vibrations of risers with nonconstant tension have then been analyzed using different approximations, since the basic equation has no analytical solution for the general case. First, the bending stiffness has been neglected ($EI = 0$), which leads to what is termed the *Bessel cable* solution, since it involves Bessel functions. The solution is exact but of little value, since Bessel functions are difficult to handle.

It has then been shown that a slight change to a term of little significance in the basic equation leads to what is termed the *simple cable* solution, involving only trigonometric functions. That solution leads to a wealth of simple expressions for all the phenomena of interest and allows many conclusions to be drawn. The modal periods are equal to twice the time required for a transverse wave to pass between adjacent nodes. The mean celerity of the whole riser is equal to the mean of the celerities at the extremities (equation [15.21]). The mean celerity between adjacent nodes is equal to the mean of the celerities at the nodes (equation [15.36]). The fundamental period (mode 1) is equal to the time for a transverse wave to descend and reascend the riser. For mode *n,* the natural period is equal to the fundamental divided by the mode number (equation [15.19]). For mode *n,* the scale of $\sqrt{\text{tension}}$ is divided into $2n$ equal intervals by all the nodes and antinodes (equation [15.31]). Maximum curvature occurs close to the lowest antinode and depends on the height of the first node, not the water depth or the riser length (equation [15.45]). The bottom-end angle depends on the bottom-end tension and the modal frequency (equation [15.39]).

Simplified analysis of risers with bending stiffness, termed *simple beam analysis,* has then been explained. In this method, the bending stiffness is treated as an increase in tension between nodes, with a different tension increase for each pair of nodes. Results obtained are in good agreement with results of the exact *numerical beam analysis*, but the solution has to be found by iteration. Bending stiffness causes all the nodes to be raised, but by different amounts.

A further simplified method of analysis called *approximate beam analysis* has been described. In this method, bending stiffness is taken into account as a mean constant increase in tension over the total length of the riser. That is better than ignoring bending stiffness completely, but results are not as accurate as those given by simple beam analysis. Results are obtained without iteration.

The Excel file *Uniform-Transverse-Modal.xls* allows all important results of five analysis methods (Bessel cable, simple cable, approximate beam, simple beam, and numerical beam) to be compared graphically and in tabular form.

The simple cable and numerical beam methods of analysis have also been adapted to risers composed of six segments with different characteristics. The Excel file *Segmented-Transverse-Modal.xls* allows the mode shapes and other results of the two methods to be compared. Large abrupt changes in segment characteristics can lead to strange mode shapes, but with good agreement between the two models.

Finally, the method used in simple cable analysis has been extended to catenary risers to obtain the natural periods of transverse vibrations (as opposed to in-plane lateral vibrations). The simple formulae have been applied to an example for which data and results have been published.

References

[1]Sparks, C. P. 2002. Transverse modal vibrations of vertical tensioned risers. *Revue de l'Institut Français du Pétrole.* 57 (1), 71–85;
Senjanović, I., A. M. Ljuština, and J. Parunov. 2006. Analytical procedure for natural vibration analysis of tensioned risers. *Journal of International Shipbuilding Progress.* Volume 53, No. 3, IOS Press, Amsterdam.

[2]Blevins, R. D. 1977. *Flow Induced Vibrations*. New York: Van Nostrand–Rienhold;
Sarpkaya, T., and M. Isaacson. 1981. *Mechanics of Wave Forces on Offshore Structures*. New York: Van Nostrand–Rienhold;
Strouhal, V. 1878. Über eine besondere Art der Tonerregung. *Annalen der Physik und Chemie,* Neue Folge, Leipzig, Germany, 216–251.

[3]Kaasen, K. E., H. Lie, F. Solaas, and J. K. Vandiver. 2000. Norwegian Deepwater Program: Analysis of vortex induced vibrations of marine risers based on full scale measurements. Paper OTC 11997, presented at the Offshore Technology Conference, Houston;
Halse, K. H. 2000. Norwegian Deepwater Program: Improved prediction of vortex induced vibrations. Paper OTC 11996, presented at the Offshore Technology Conference, Houston;
Cornut, S. F. A., and J. K. Vandiver. 2000. Offshore VIV monitoring at Schiehallion—analysis of riser VIV response. Paper OMAE 2000/Pipe-5022, presented at the Offshore Mechanics and Arctic Engineering Conference, New Orleans.

[4]Timoshenko, S. 1955. *Vibration Problems in Engineering*. New York: Van Nostrand.

[5]Senjanović, Ljuština, and Parunov, 2006.

[6]Sparks, 2002; Senjanović, Ljuština, and Parunov, 2006.

[7]Croutelle, Y, C. Ricbourg, C. Le Cunff, J.-M. Heurtier, and F. Biolley. 2002. Fatigue of steel catenary risers. Paper presented at the Ultra Deep Engineering and Technology Conference, Brest, France.

Appendix A: Tensioned-Beam Equations

The differential equations governing the static deflected shape of a tensioned beam can be deduced very simply by considering the balance of forces acting on a short segment. The cases of small- and large-angle deflections need to be considered separately.

Small-Angle Deflections

Figure A–1 shows a short segment of tensioned beam, orientated close enough to the vertical (within 10°) for small-angle deflection theory to be applicable.

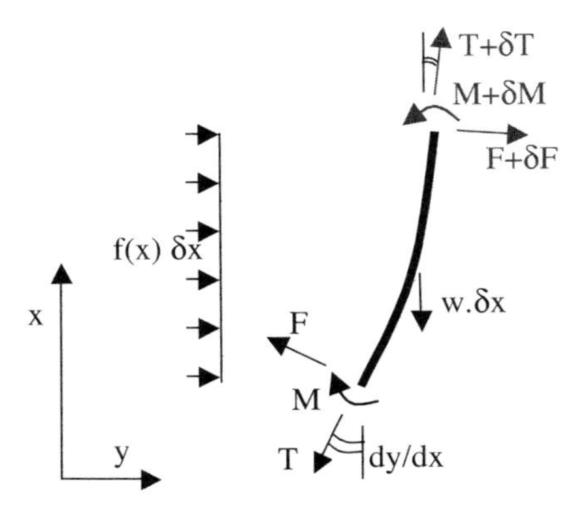

Fig. A–1. *Tensioned beam—forces acting on a segment of length* δx

Considering static forces acting on a beam segment of length δx and resolving in the horizontal direction gives

$$\delta F + \delta\left(T\frac{dy}{dx}\right) + f(x)\delta x = 0 \qquad (A.1)$$

where F is the shear force, $T\,dy/dx$ is the horizontal component of axial tension, and $f(x)$ is the external lateral load. Hence,

$$\frac{dF}{dx} + \frac{d}{dx}\left(T\frac{dy}{dx}\right) + f(x) = 0 \qquad (A.2)$$

However, the shear force F is related to the moment by $F = dM/dx$. Hence,

$$\frac{d^2M}{dx^2} + \frac{d}{dx}\left(T\frac{dy}{dx}\right) + f(x) = 0 \qquad (A.3)$$

Equation (A.3) applies to all tensioned beams with small-angle deflections and is the most general form of the governing equation. The x-axis is shown vertical, but equation (A.3) is also valid for nonvertical axes, providing that the component of the gravitational load, normal to the beam axis, is included in $f(x)$.

For the special case of a beam made of elastic materials with bending stiffness EI, $M = -EI\,d^2y/dx^2$. Hence,

$$\frac{d^2}{dx^2}\left(EI\frac{d^2y}{dx^2}\right) - \frac{d}{dx}\left(T\frac{dy}{dx}\right) - f(x) = 0 \qquad (A.4)$$

For the case of a tensioned beam with constant EI, equation (A.4) becomes

$$EI\frac{d^4y}{dx^4} - \frac{d}{dx}\left(T\frac{dy}{dx}\right) - f(x) = 0 \qquad (A.5)$$

For near-vertical tensioned beams, $dT/dx = w$, the weight per unit length. Equation (A.5) can then be rewritten as

$$EI\frac{d^4y}{dx^4} - T\frac{d^2y}{dx^2} - w\frac{dy}{dx} - f(x) = 0 \qquad (A.6)$$

Equation (A.6) is the form most frequently used for near-vertical risers made of elastic materials with constant bending stiffness. Note, however, that for such risers, equations (A.3)–(A.6) are identical.

Large-Angle Deflections

For the case of large-angle deflections, it is simplest to rederive the basic equation from the force system acting on a short segment of tensioned beam, as shown in figure A–2.

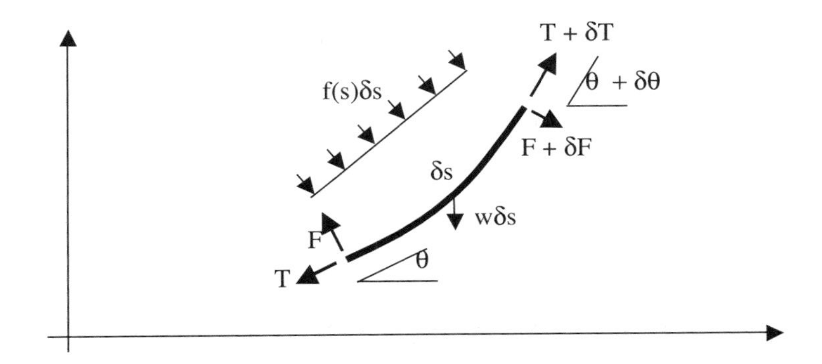

Fig. A–2. *Large-angle deflections—forces acting on a beam segment of length δs*

Resolving forces normal to the segment axis leads to

$$\delta F - T\ \delta\theta + w\ \delta s \cos\theta + f(s)\ \delta s = 0 \qquad (A.7)$$

where F is the shear force and $f(s)$ is the external load. Equation (A.7) then leads to

$$\frac{dF}{ds} - T\frac{d\theta}{ds} + w\cos\theta + f(s) = 0 \qquad (A.8)$$

However, the shear force $F = dM/ds$; hence,

$$\frac{d^2M}{ds^2} - T\frac{d\theta}{ds} + w\cos\theta + f(s) = 0 \qquad (A.9)$$

Further, the curvature $1/R = M/EI = d\theta/ds$; hence,

$$\frac{d^2}{ds^2}\left(EI\frac{d\theta}{ds}\right) - T\frac{d\theta}{ds} + w\cos\theta + f(s) = 0 \qquad (A.10)$$

Therefore, for the case of constant EI,

$$EI\frac{d^3\theta}{ds^3}-T\frac{d\theta}{ds}+w\cos\theta+f(s)=0 \tag{A.11}$$

Equation (A.11) is the form most frequently used for catenary risers made of elastic materials with constant bending stiffness. Note, however, that for such risers, equations (A.9)–(A.11) are identical.

Resolution of forces in the axial direction leads to equation (A.12), which relates the axial tension to the beam angle θ:

$$\frac{dT}{ds}=w\sin\theta \tag{A.12}$$

Comments and Reminders

Note that the preceding equations are independent of the axial stiffness EA. Axial stretch is a second-order effect.

In the derivation of the preceding equations, the relationships $F = dM/ds$, $M = EI/R$, and $1/R = d\theta/ds$ were used and can be found in standard textbooks on strength of materials. Some readers may nevertheless appreciate a brief reminder as to how they are obtained, as well as a short discussion about convergence between the small- and large-angle deflection equations.

Figure A–3*a* shows the elevation of a short length δs of beam subject to shear force F and moment M. The equilibrium of moments leads to Equation A.13 and hence to Equation A.14:

$$\delta M = F\,\delta s \tag{A.13}$$

$$F=\frac{dM}{ds} \tag{A.14}$$

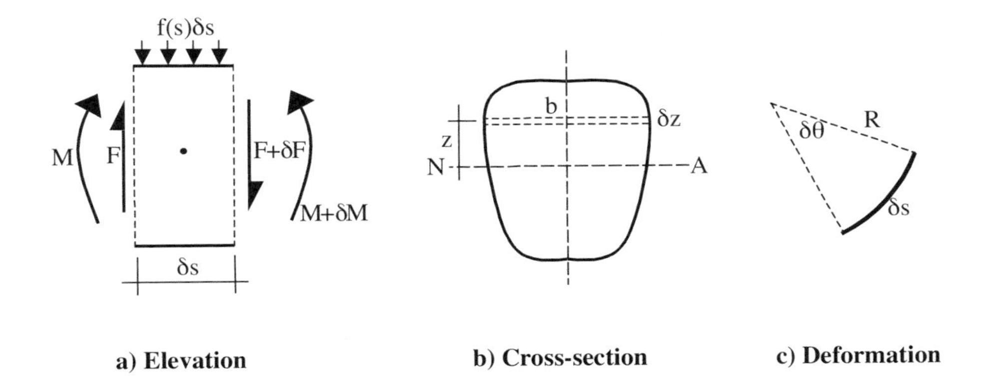

Fig. A–3. *Beam sketches*

Figure A–3*b* shows a beam cross-section of any shape, symmetrical about the vertical axis. Assuming that plane sections remain plane, if the beam is bent in the vertical plane to radius R, then the axial strain at distance z from the neutral axis NA will be equal to z/R, with corresponding stress Ez/R, where E is the Young's modulus of the material. The stress acting on the strip of width b, of height δz, and at distance z from the neutral axis NA induces moment $\delta M = bz^2\, \delta z/R$ about the neutral axis NA. Integration over the complete cross-section then yields

$$M = \frac{EI}{R} \qquad (A.15)$$

where I is the second moment of area.

Figure A–3*c* shows the global deformation of a length δs of beam, bent to radius R. The angle $\delta\theta$ turned through is related to δs by $\delta s = R\,\delta\theta$. Hence,

$$\frac{1}{R} = \frac{d\theta}{ds} \qquad (A.16)$$

Convergence between Small- and Large-Angle Deflection Equations

The small-angle deflection equations derived in the first section apply to beams close to the vertical x-axis (see fig. A–1). By contrast, the large-angle deflection equations are expressed in terms of the angle θ with the horizontal (see fig. A–2). If, instead, the latter are expressed in terms of the angle ψ with the vertical, where $\psi = (\pi/2) - \theta$, then the small- and large-angle deflection equations should converge for small values of ψ.

If $\psi = (\pi/2) - \theta$, then $\delta\psi = -\delta\theta$, and $\cos\theta = \sin\psi$. Hence, equation (A.11) can be rewritten as follows:

$$-EI\frac{d^3\psi}{ds^3}+T\frac{d\psi}{ds}+w\sin\psi+f(s)=0 \qquad (A.17)$$

Small-angle deflection theory assumes that the angle ψ between the beam axis and the reference axis is everywhere small enough for the following approximations to be valid: $\sin\psi = \psi = dy/dx$, and $\delta x = \delta s$. With those approximations, equation (A.17) becomes equation (A.18), which is the same as equation (A.6):

$$-EI\frac{d^4y}{dx^4}+T\frac{d^2y}{dx^2}+w\frac{dy}{dx}+f(x)=0 \qquad (A.18)$$

The accuracy of the small-angle deflection equations decreases as ψ increases. Small-angle deflection theory is generally considered to be valid for angles up to 10°.

Appendix B: Tension Calculations for Simple Riser Cases

This appendix shows how the bullet points in the final section of chapter 2 can be applied to obtain the effective tension T_e for the five vertical uniform tubes shown in figure B–1.

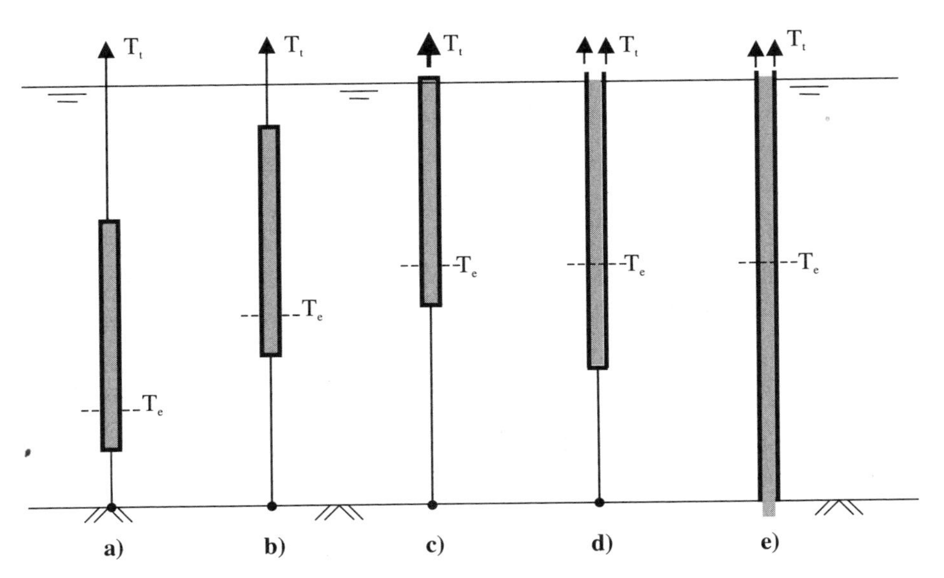

Fig. B–1. *Five vertical uniform tubes*

The tubes of figure B–1 have identical cross-sectional areas (A_i and A_e). They are filled with fluids of the same density and are subject to the same top tensions T_t. They also all have the same upper-segment lengths, between the top end and the "section of interest" (shown by dotted lines in fig. B–1). The five upper segments therefore all have the same apparent weight W_a.

From the bullet point following equation (2.19), if the top tensions T_t and the upper-segment apparent weights W_a are all equal, then the effective tensions T_e must also all be equal ($T_e = T_t - W_a$) at the section of interest. However, the pressures acting in and on tubes *a* through *d* are all different. Hence, from equation (2.19), the true wall tensions T_{tw} at the section level must all be different.

For such simple examples, the true wall tension T_{tw} at the section level can be calculated by considering all the vertical loads transmitted to the tube wall, including the pressure loads acting on the top-end cap, which are zero for tubes *d* and *e*. The true wall tension T_{tw} at the section level is then given by equation (B.1), where W_{tube} is the weight of the tube upper segment (tube wall *in air*) and $T_{tw\ top}$ is the tension in the tube wall, just below the top-end cap.

$$T_{tw\,section} = T_{tw\ top} - W_{tube} \qquad (B.1)$$

The wall tension $T_{tw\ top}$ below the top-end cap can be expressed in terms of the top tension and the pressures acting on the top-end cap as:

$$T_{tw\,top} = T_t + (p_i A_i - p_e A_e)_{top} \qquad (B.2)$$

The vertical equilibrium of the upper-segment internal fluid column (true weight W_i) requires that

$$(p_i A_i)_{top} = (p_i A_i)_{section} - W_i \qquad (B.3)$$

Likewise, the equilibrium of the displaced fluid column (true weight W_e) requires that

$$(p_e A_e)_{top} = (p_e A_e)_{section} - W_e \qquad (B.4)$$

Hence, from equations (B.2)–(B.4),

$$T_{tw\,top} = T_t + (p_i A_i - p_e A_e)_{section} - W_i + W_e \qquad (B.5)$$

and from equation (B.1),

$$T_{\text{tw section}} = T_{\text{t}} + (p_{\text{i}}A_{\text{i}} - p_{\text{e}}A_{\text{e}})_{\text{section}} - W_{\text{i}} + W_{\text{e}} - W_{\text{tube}} \qquad (B.6)$$

However, $W_{\text{tube}} + W_{\text{i}} - W_{\text{e}} = W_{\text{a}}$, which is the apparent weight. Hence,

$$T_{\text{tw section}} = T_{\text{t}} - W_{\text{a}} + (p_{\text{i}}A_{\text{i}} - p_{\text{e}}A_{\text{e}})_{\text{section}} \qquad (B.7)$$

Still, from equation (2.19), at the section of interest,

$$T_{\text{e section}} = (T_{\text{tw}} - p_{\text{i}}A_{\text{i}} + p_{\text{e}}A_{\text{e}})_{\text{section}} \qquad (B.8)$$

Thus, substitution into equation (B.7) gives

$$T_{\text{e section}} = T_{\text{t}} - W_{\text{a}} \qquad (B.9)$$

Equation (B.9) is coherent with the definition of effective tension T_{e} given by the bullet point following equation (2.19), which allows the effective tension to be written down at sight. Equation (B.8) can then be used to work back to the true wall tension T_{tw}.

For all but the simplest of cases, such as the five above, it is always simplest to find the effective tension *before* the true wall tension. End conditions (capped or not) do not influence equation (B.8) or (B.9).

In the cases of tubes *d* and *e*, note that a change in internal pressure can be obtained only by changing the internal fluid (level or density), which changes the segment apparent weight W_{a}.

Appendix C: Application of the Morison Equation to Risers

Morison derived his equation while studying wave-induced hydrodynamic forces on piles.[1] Although the equation has been considered controversial for many years, it has nevertheless been used almost exclusively to calculate the combined effects of current and wave loads on cylindrical structures such as risers, since its first formulation in 1950.[2] It is considered to give the forces on circular cylinders with reasonable accuracy, providing that the diameter of the cylinder is small compared to the wavelength.

Morison applied strip theory to calculate the force per unit length of pile, in two dimensions. He decomposed the hydrodynamic force f_H, acting normal to the pile, into two components—a drag force f_D, resulting from the velocity of the flow past the body, plus an inertia force f_I, due to the acceleration of the flow:

$$f_H = f_D + f_I \qquad (C.1)$$

The Drag Force

The drag force f_D has been thoroughly investigated in the laboratory for the case of steady flow and has been found to vary with the square of the velocity.[3] For circular cylinders, exposed to flow normal to their axis, the force per unit length is given by:

$$f_D = \frac{1}{2}\rho C_D \phi u|u| \qquad (C.2)$$

where ρ is the fluid density, C_D is the nondimensional drag coefficient, ϕ is the body diameter, and u is the instantaneous velocity of the fluid (i.e., the velocity in the absence of the cylinder) normal to the cylinder axis. C_D varies with the body shape and the Reynolds number, but for bare circular cylinders, it typically has a value of about 1.0 for laminar flow (subcritical flow) and about 0.6–0.7 for turbulent flow (supercritical flow).

For a riser that is itself moving laterally with velocity v in the direction of the flow, the relative velocity must be used in equation (C.2), which then becomes

$$f_D = \frac{1}{2}\rho C_D \phi (u - v)|u - v| \qquad (C.3)$$

The Inertia Force

The inertia force f_I due to fluid acceleration has also been extensively investigated in the laboratory.[4] For a volume V of fluid of density ρ undergoing a uniform acceleration $\dot{u}$, the dynamic pressure field acting on it must apply an inertia force f_I given by:

$$f_I = \rho V \dot{u} \qquad (C.4)$$

For a stationary sphere of volume V subjected to accelerating flow, the inertia force has been found in the laboratory to be given by

$$f_I = C_M \rho V \dot{u} \qquad (C.5)$$

where C_M is the nondimensional inertia coefficient and $\dot{u}$ is the instantaneous acceleration of the fluid (in the absence of the sphere).

Equation (C.5) can be decomposed into two parts: the hydrodynamic force acting on the displaced fluid in the absence of the sphere ($\rho V \dot{u}$), plus an additional force $(C_M - 1)\rho V \dot{u}$ due to the acceleration of the fluid relative to the sphere. If the sphere itself is moving with acceleration $\dot{v}$ in the same direction as the fluid, then the relative acceleration becomes $\dot{u} - \dot{v}$ and the inertia force becomes

$$f_I = \rho V \dot{u} + (C_M - 1)\rho V(\dot{u} - \dot{v}) \qquad (C.6)$$

The inertia force given by equation (C.6) varies linearly with $\dot{v}$. It agrees with equation (C.4) for a sphere accelerating precisely with the fluid ($\dot{v} = \dot{u}$) and with equation (C.5) for the case of a stationary sphere ($\dot{v} = 0$). Equation (C.6) can be rewritten as

$$f_I = C_M \rho V \dot{u} - (C_M - 1)\rho V \dot{v} \qquad (C.7)$$

The term $(C_M - 1)\rho V$ is frequently called the *added mass* since it has units of mass and shares the same acceleration $\dot{v}$ as the sphere itself.

Equations (C.6) and (C.7) apply to spheres in uniformly accelerated flow. They can also be used to give the inertia force per unit length of a small-diameter circular cylinder, such as a riser subject to wave action, if V is replaced by the external cross-sectional area A_e. Hence, for a riser, Morison's equation for the hydrodynamic force per unit length (eq. [C.1]) can be written as either of the following:

$$f_H = \frac{1}{2}\rho C_D \phi (u - v)|u - v| + \rho A_e \dot{u} + (C_M - 1)\rho A_e (\dot{u} - \dot{v}) \qquad (C.8)$$

$$f_H = \frac{1}{2}\rho C_D \phi (u - v)|u - v| + C_M \rho A_e \dot{u} - (C_M - 1)\rho A_e \dot{v} \qquad (C.9)$$

For a smooth cylinder, at high Reynolds number, the inertia coefficient is typically close to $C_M = 2$.

The Controversy

The Morison equation is controversial principally because the drag force term is nonlinear. When the flow is not perpendicular to the riser axis, which is generally the case, there are two ways to proceed: The drag force can be evaluated in the direction of the flow and then resolved into components perpendicular and parallel to the riser axis; alternatively, the flow velocity can be resolved into components perpendicular and parallel to the riser axis, and the force components can then be evaluated. These approaches do not give the same result. Today, however, it is general practice to resolve the velocity perpendicular to the riser axis and neglect the effect of the component parallel to the axis.

When the *acceleration* vectors of the fluid and the riser are in different directions, the inertia force can be obtained without difficulty by simple resolution of the vectors into two mutually perpendicular directions. Evaluation of the nonlinear drag force, when the *velocity* vectors are not aligned, is not so simple! Different methods can be used. One method consists of resolving the term (u - v) into two mutually perpendicular directions (e.g., x and y) while replacing $|u-v|$ by the modulus of the true resultant relative velocity vector $|v_a|^5$. The x and y components of the drag force per unit length then become

$$f_{D_x} = \frac{1}{2}\rho C_D \phi (u_x - v_x)|v_a| \qquad (C.10)$$

$$f_{D_y} = \frac{1}{2}\rho C_D \phi (u_y - v_y)|v_a| \qquad (C.11)$$

where

$$v_a = \sqrt{(u_x - v_x)^2 + (u_y - v_y)^2} \qquad (C.12)$$

The nonlinear nature of the drag force poses further problems for frequency domain analyses. There, the general practice is to replace the nonlinear drag force with a linearized force that dissipates the same quantity of energy.

Decrease of Wave-Induced Drag and Inertia Forces with Depth

Expressions for the fluid velocities and accelerations below a wave vary according to the wave theory used. The Airy wave theory is the simplest, although it only applies to small-amplitude waves. It gives the horizontal and vertical velocities as follows:

$$u_h = \frac{H}{2}\omega \frac{\cosh \lambda(d-z)}{\sinh \lambda d} \cos \omega t \qquad (C.10)$$

$$u_v = \frac{H}{2}\omega \frac{\sinh \lambda(d-z)}{\sinh \lambda d} \sin \omega t \qquad (C.11)$$

$$\dot{u}_h = -\frac{H}{2}\omega^2 \frac{\cosh\lambda(d-z)}{\sinh\lambda d}\sin\omega t \qquad (C.12)$$

$$\dot{u}_v = \frac{H}{2}\omega^2 \frac{\sinh\lambda(d-z)}{\sinh\lambda d}\cos\omega t \qquad (C.13)$$

where subscripts "h" and "v" denote the horizontal and vertical components of fluid velocity and acceleration at depth z; H is the wave (double) amplitude, ω is the circular frequency (equal to $2\pi/T_p$ for wave period T_p), $\lambda = 2\pi/L$ for wavelength L, and d is the total depth.

The wave-induced velocities and accelerations both decrease approximately exponentially with depth (with $e^{-\lambda z}$). The drag force, being a function of velocity squared, decreases more rapidly than the inertia force.

Complicated Riser Geometries

For riser geometries more complicated than bare pipe, such as drilling risers equipped with kill and choke lines, an equivalent diameter ϕ and an equivalent cross-sectional area A_e must be used in the Morison equation. Likewise, the coefficients C_D and C_M have to be adapted to the geometry and state of a riser (e.g., with or without marine growth). A great deal of work was done in the late 1970s to determine the values of the equivalent diameters, sections, and coefficients for various riser geometries.[6]

Influence of Vortex-Induced Vibrations

In the case of transverse oscillations resulting from vortex-induced vibrations (VIV), the in-line drag force can increase significantly.[7] It is as though the riser diameter in equation (C.3) should be replaced by a larger diameter, close to the *swept double amplitude* of the riser transverse oscillation. Laboratory experiments on elastically-supported cylinders have shown that VIV oscillations are auto-limited to a (single) amplitude of about one diameter. It is thought probable that riser oscillations are subject to the same limit. Nevertheless, given the shortage of real riser VIV data, this limit cannot be definitely affirmed. VIV is the subject of intensive ongoing research.

References

[1]Morison, J. R., M. P. O'Brien, J. W. Johnson, and S. A. Shaff. 1950. The forces exerted by surface waves on piles. *Petroleum Transactions, AIME* 189, 149–154

[2]Wade, B. G., and M. Dwyer. 1976. On the application of Morison's equation to fixed offshore platforms. Paper OTC 2723, presented at the Offshore Technology Conference, Houston.

[3]Newman, J. N. 1977. *Marine Hydrodynamics*. Cambridge, MA: MIT Press; Chakrabarti, S. K. 1984. Moored floating structures and hydrodynamic coefficients. Paper presented at the Ocean Structural Dynamics Symposium, Corvallis, OR.

[4]Sarpkaya, T., and M. Isaacson. 1981. *Mechanics of Wave Forces on Offshore Structures*. New York: Van Nostrand–Rienhold;

Chakrabarti, S. K. 1987. *Hydrodynamics of Offshore Structures*. Computational Mechanics Publications, WIT Press, Southampton, UK.

[5]Molin, B. 2002. *Hydrodynamique des Structures Offshore*. Editions Technip, Paris, France.

[6]Ottsen Hansen, N.-E. 1979. Hydrodynamic forces on composite risers and individual cylinders. Paper OTC 3541, presented at the Offshore Technology Conference, Houston.

[7]Blevins, R. D. 1977. *Flow Induced Vibrations*. New York: Van Nostrand–Rienhold;
Faltinsen, O. M. 1990. *Sea Loads on Ships and Offshore Structures*. Cambridge University Press, Cambridge, UK.

Appendix D: Stress and Strain Relationships in a Thick-Walled Pipe

General Stress Relationships

Figure D–1*a* shows the circumferential and radial stresses acting on a small segment within a pipe wall.

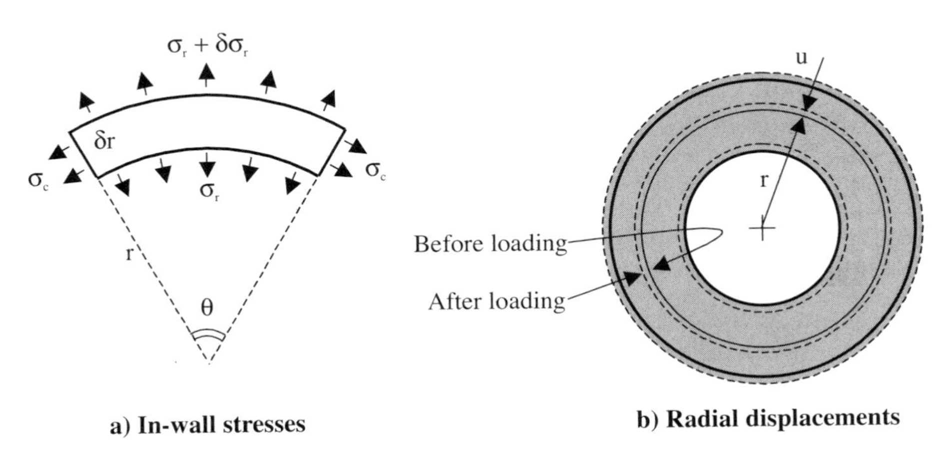

Fig. D–1. *In-wall pressure-induced stresses and radial displacements*

Resolving forces in the radial direction yields

$$\sigma_c\ \delta r\theta = (\sigma_r + \delta\sigma_r)(r + \delta r)\theta - \sigma_r r\theta \qquad (D.1)$$

Thus,

$$\sigma_c \, \delta r = r \, \delta\sigma_r + \sigma_r \, \delta r \qquad (D.2)$$

Equation (D.2) leads to equations (D.3)–(D.5):

$$\sigma_c = \frac{d}{dr} r\sigma_r \qquad (D.3)$$

$$\sigma_c - \sigma_r = r\frac{d}{dr}\sigma_r \qquad (D.4)$$

$$\sigma_c + \sigma_r = \frac{1}{r}\frac{d}{dr} r^2\sigma_r \qquad (D.5)$$

From equation (D.5), the integral of the mean stress $(\sigma_c+\sigma_r)$ / 2 across the pipe cross-section gives

$$\int_{r_i}^{r_e} \left(\frac{\sigma_c + \sigma_r}{2}\right) 2\pi r \, dr = \pi \int_{r_i}^{r_e} d(r^2\sigma_r) = p_i A_i - p_e A_e \qquad (D.6)$$

where r_i and r_e are the internal and external radii. No assumptions about material properties were made in deriving equation (D.6). Hence, the integral of $(\sigma_c+\sigma_r)$ / 2 across the pipe cross-section is always equal to $(p_i A_i - p_e A_e)$.

The special characteristic of elastic isotropic tubes is that $(\sigma_c+\sigma_r)$ / 2 is constant at all points of the wall section, as given by

$$\frac{\sigma_c + \sigma_r}{2} = \frac{p_i A_i - p_e A_e}{A_e - A_i} \qquad (D.7)$$

This was proved by Lamé, as shown by equations (D.18)-(D.35). Note that equation (D.7) is also satisfied by the mean values of circumferential stress $\overline{\sigma_c}$ and radial stress $\overline{\sigma_r}$. Those mean stresses are given by

$$\overline{\sigma_c} = \frac{p_i r_i - p_e r_e}{r_e - r_i} \qquad (D.8)$$

$$\overline{\sigma_r} = -\frac{p_i r_i + p_e r_e}{r_e + r_i} \qquad (D.9)$$

Strain Relationships for Thick-Walled Pipes

The relationship between the circumferential strain ε_c and the radial strain ε_r is also independent of material properties. Figure D–1*b* shows the small radial displacement *u* of a cylindrical surface in the pipe wall under pressure loading. The circumferential and radial strains ε_c and ε_r are then given by

$$\varepsilon_c = \frac{u}{r} \qquad (D.10)$$

$$\varepsilon_r = \frac{du}{dr} \qquad (D.11)$$

Equations (D.12)–(D.14) are straightforward mathematical relationships between two variables: *u* and *r*.

$$\frac{du}{dr} = \frac{d}{dr}\left(r\frac{u}{r}\right) \qquad (D.12)$$

$$\frac{du}{dr} - \frac{u}{r} = r\frac{d}{dr}\left(\frac{u}{r}\right) \qquad (D.13)$$

$$\frac{du}{dr} + \frac{u}{r} = \frac{1}{r}\frac{d}{dr}\left(r^2\frac{u}{r}\right) \qquad (D.14)$$

By use of the substitution from equations (D.10) and (D.11), equations (D12)–(D.14) yield the three strain relationships of equations (D.15)–(D.17).

$$\varepsilon_r = \frac{d}{dr}(r\varepsilon_c) \qquad (D.15)$$

$$\varepsilon_r - \varepsilon_c = r\frac{d}{dr}(\varepsilon_c) \qquad (D.16)$$

$$\varepsilon_r + \varepsilon_c = \frac{1}{r}\frac{d}{dr}\left(r^2\varepsilon_c\right) \qquad (D.17)$$

Equation (D.15)–(D.17) parallel the three stress relationships of equations (D.3)–(D.5):

Stress-Strain Equations for Thick-Walled Pipes

Lamé's equations for stresses and strains in thick-walled circular cylindrical pipes, made from elastic isotropic materials, can be found in many textbooks. They are rederived here for completeness.

The principal wall stresses σ_a, σ_c, and σ_r and strains ε_a, ε_c, and ε_r are related by the classical equation (D.18), which, when inverted, gives equation (D.19):

$$\begin{vmatrix} \varepsilon_a \\ \varepsilon_c \\ \varepsilon_r \end{vmatrix} = \frac{1}{E} \begin{vmatrix} 1 & -\nu & -\nu \\ -\nu & 1 & -\nu \\ -\nu & -\nu & 1 \end{vmatrix} \begin{vmatrix} \sigma_a \\ \sigma_c \\ \sigma_r \end{vmatrix} \qquad (D.\ 18)$$

$$\begin{vmatrix} \sigma_a \\ \sigma_c \\ \sigma_r \end{vmatrix} = \frac{E}{(1+\nu)(1-2\nu)} \begin{vmatrix} (1-\nu) & \nu & \nu \\ \nu & (1-\nu) & \nu \\ \nu & \nu & (1-\nu) \end{vmatrix} \begin{vmatrix} \varepsilon_a \\ \varepsilon_c \\ \varepsilon_r \end{vmatrix} \qquad (D.\ 19)$$

where E is Young's modulus and ν is Poisson's ratio. Hence, from the lower two lines of equation (D. 19) and equations (D.10) and (D.11),

$$\sigma_c - \sigma_r = \frac{E}{1+\nu}(\varepsilon_c - \varepsilon_r) = \frac{E}{1+\nu}\left(\frac{u}{r} - \frac{du}{dr}\right) \qquad (D.20)$$

Differentiating the bottom line of equation (D. 19) and multiplying by r gives

$$r\frac{d}{dr}(\sigma_r) = \frac{E}{(1+\nu)(1-2\nu)}\left[\nu r\frac{d\varepsilon_a}{dr} + \nu r\frac{d\varepsilon_c}{dr} + (1-\nu)r\frac{d\varepsilon_r}{dr}\right] \qquad (D.21)$$

Equation (D.16), together with differentiation of equation (D.11), leads to equations (D.22) and (D.23):

$$r\frac{d\varepsilon_c}{dr} = \frac{du}{dr} - \frac{u}{r} \qquad (D.22)$$

$$\frac{d\varepsilon_r}{dr} = \frac{d^2u}{dr^2} \tag{D.23}$$

Substituting equations (D.22) and (D.23) into equation (D.21) and assuming plane sections remain plane (i.e., $d\varepsilon_a/dr = 0$) gives

$$r\frac{d}{dr}(\sigma_r) = \frac{E}{(1+\nu)(1-2\nu)}\left[\nu\left(\frac{du}{dr} - \frac{u}{r}\right) + (1-\nu)r\frac{d^2u}{dr^2}\right] \tag{D.24}$$

However, from equations (D.4) and (D.20),

$$r\frac{d}{dr}(\sigma_r) = \sigma_c - \sigma_r = \frac{E}{1+\nu}\left(\frac{u}{r} - \frac{du}{dr}\right) \tag{D.25}$$

Hence, from equations (D.24) and (D.25),

$$\frac{1}{1-2\nu}\left[\nu\left(\frac{du}{dr} - \frac{u}{r}\right) + (1-\nu)r\frac{d^2u}{dr^2}\right] = \frac{u}{r} - \frac{du}{dr} \tag{D.26}$$

which simplifies to

$$\frac{du}{dr} - \frac{u}{r} + r\frac{d^2u}{dr^2} = 0 \tag{D.27}$$

with the following solution:

$$u = Ar + \frac{B}{r} \tag{D.28}$$

Hence,

$$\varepsilon_c = \frac{u}{r} = A + \frac{B}{r^2} \tag{D.29}$$

and

$$\varepsilon_r = \frac{du}{dr} = A - \frac{B}{r^2} \tag{D.30}$$

Substitution of equations (D.29) and (D.30) into the lower two rows of equation (D.19) (and the limit conditions on the inner and outer surfaces: $\sigma_{r_i} = -p_i$, and $\sigma_{r_e} = -p_e$) leads to equations (D.31) and (D.32):

$$\sigma_c = \frac{p_i r_i^2 - p_e r_e^2}{r_e^2 - r_i^2} + \frac{(p_i - p_e) r_i^2 r_e^2}{\left(r_e^2 - r_i^2\right) r^2} \tag{D.31}$$

$$\sigma_r = \frac{p_i r_i^2 - p_e r_e^2}{r_e^2 - r_i^2} - \frac{(p_i - p_e) r_i^2 r_e^2}{\left(r_e^2 - r_i^2\right) r^2} \tag{D.32}$$

where r_i, r_e, and r are the pipe radii (internal radius, external radius, and radius of interest; see fig. D–1). Putting $A_i = \pi r_i^2$, $A_e = \pi r_e^2$, and $A_r = \pi r^2$ gives

$$\sigma_c = \frac{p_i A_i - p_e A_e}{A_e - A_i} + \frac{(p_i - p_e) A_i A_e}{(A_e - A_i) A_r} \tag{D.33}$$

$$\sigma_r = \frac{p_i A_i - p_e A_e}{A_e - A_i} - \frac{(p_i - p_e) A_i A_e}{(A_e - A_i) A_r} \tag{D.34}$$

Hence,

$$\frac{\sigma_c + \sigma_r}{2} = \frac{p_i A_i - p_e A_e}{A_e - A_i} \tag{D.35}$$

$$\frac{\sigma_c - \sigma_r}{2} = \frac{(p_i - p_e) A_i A_e}{(A_e - A_i) A_r} \tag{D.36}$$

Equations (D.35) and (D.36) are coherent with equations (4.5), (4.6), (4.8), and (4.9) of chapter 4.

Axial Stress for Thick-Walled Pipes

From the first row of equation (D.18), the axial strain is given by

$$\varepsilon_a = \frac{1}{E}(\sigma_a - \nu\sigma_c - \nu\sigma_r) \tag{D.37}$$

However, the sum of the stresses ($\sigma_c + \sigma_r$) is constant from equation (D.35). Hence, if plane sections remain plane (i.e., ε_a is constant), then the axial stress σ_a is also constant across the section.

Appendix E: Equivalent Poisson's Ratios for Anisotropic Pipes

Ratios Deduced from Material Characteristics

The equivalent Poisson's ratios $\overline{\nu_i}$ and $\overline{\nu_e}$, defined in chapter 5 for anisotropic pipes, can be derived from the Poisson's ratios of the basic anisotropic material. For an orthotropic material, such as a balanced composite laminate, with its proper axes aligned with the pipe axes, there is no interrelationship between normal and shear stresses, in the absence of torsion. The three dimensional in-wall stress-strain equation for a pipe made of such material is given by

$$\begin{vmatrix} \varepsilon_a \\ \varepsilon_c \\ \varepsilon_r \end{vmatrix} = \begin{vmatrix} 1/E_a & -\nu_{ac}/E_c & -\nu_{ar}/E_r \\ -\nu_{ca}/E_a & 1/E_c & -\nu_{cr}/E_r \\ -\nu_{ra}/E_a & -\nu_{rc}/E_c & 1/E_r \end{vmatrix} \begin{vmatrix} \sigma_a \\ \sigma_c \\ \sigma_r \end{vmatrix} \tag{E.1}$$

where σ_a, σ_c, and σ_r are the respective axial, circumferential, and radial stresses; ε_a, ε_c, ε_r are the corresponding strains; E_a, E_c, E_r are the corresponding Young's moduli; and ν_{ac}, ν_{ar}, ν_{ca}, ν_{cr}, ν_{ra}, ν_{rc} are the six Poisson's ratios. From the first line of equation (E.1), the axial strain is given by

$$\varepsilon_a = \frac{\sigma_a}{E_a} - \frac{\nu_{ac}}{E_c}\sigma_c - \frac{\nu_{ar}}{E_r}\sigma_r \tag{E.2}$$

Since plane sections remain plane, the axial strain ε_a is uniform across the section. Noting from the reciprocal theorem that $\nu_{ac} / E_c = \nu_{ca} / E_a$, $\nu_{cr} / E_r = \nu_{rc} / E_c$, and $\nu_{ar} / E_r = \nu_{ra} / E_a$ allows equation (E.2) to be rewritten in terms of the mean pressure-induced stresses in the pipe wall ($\overline{\sigma_a}$, $\overline{\sigma_c}$, and $\overline{\sigma_r}$), as follows:

$$\varepsilon_a = \frac{1}{E_a}\left(\overline{\sigma_a} - \nu_{ca}\overline{\sigma_c} - \nu_{ra}\overline{\sigma_r}\right) \tag{E.3}$$

However, the mean pressure-induced stresses ($\overline{\sigma_a}$, $\overline{\sigma_c}$, and $\overline{\sigma_r}$) are given in terms of the internal and external pressures p_i and p_e and the internal and external radii r_i and r_e by:

$$\overline{\sigma_a} = \frac{p_i r_i^2 - p_e r_e^2}{r_e^2 - r_i^2} \tag{E.4}$$

$$\overline{\sigma_c} = \frac{p_i r_i - p_e r_e}{r_e - r_i} \tag{E.5}$$

$$\overline{\sigma_r} = -\frac{p_i r_i + p_e r_e}{r_e + r_i} \tag{E.6}$$

For the case of *internal* pressure only, substitution of equations (E.4)–(E.6) into equation (E.3) gives

$$\varepsilon_a = \frac{\overline{\sigma_a}}{E_a}\left[1 - \nu_{ca}\left(1 + \frac{r_e}{r_i}\right) + \nu_{ra}\left(\frac{r_e}{r_i} - 1\right)\right] \tag{E.7}$$

However, axial stresses and strains, induced by internal pressure, are related by $\varepsilon_a = \left(1 - 2\overline{\nu_i}\right)\left(\overline{\sigma_a} / E_a\right)$ (cf. eq. [5.7] of chap. 5); hence,

$$1 - 2\overline{\nu_i} = 1 - \nu_{ca}\left(1 + \frac{r_e}{r_i}\right) + \nu_{ra}\left(\frac{r_e}{r_i} - 1\right) \tag{E.8}$$

Similarly, by consideration of axial stresses and strains induced by *external* pressure only, it can be deduced that

$$1 - 2\overline{\nu_e} = 1 - \nu_{ca}\left(1 + \frac{r_i}{r_e}\right) - \nu_{ra}\left(1 - \frac{r_i}{r_e}\right) \tag{E.9}$$

For isotropic materials, $\overline{\nu_i} = \overline{\nu_e} = \nu_{ca} = \nu_{ra} = \nu$.

Determination of $\overline{\nu_e}$ from an Axial Load Test

The object of this section is to show how the equivalent Poisson's ratio for external pressure $\overline{\nu_e}$ can be deduced from axial and circumferential strains induced by a *pure axial load.* Figure E–1 shows a capped uniform tube subject to external pressure p_e.

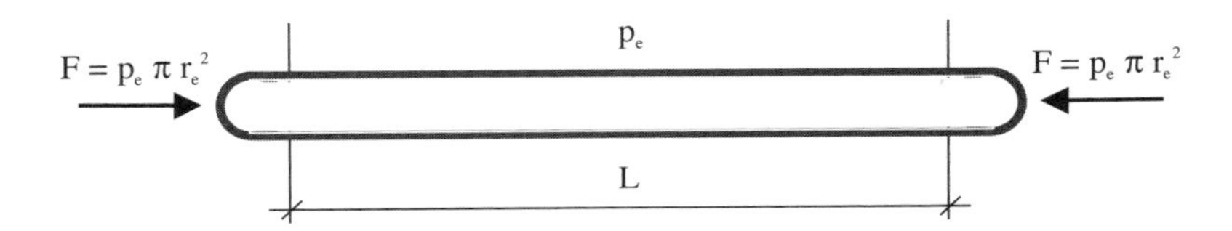

Fig. E–1. *Capped tube subject to external pressure*

The tube is subject to an *axial compressive force* ($F = p_e A_e = p_e \pi r_e^2$). The test length L is subject to a *radial force* ($p_e 2\pi r_e L = 2FL/r_e$) acting on the outer surface. The axial strain $\varepsilon_{a\,pe}$ of the test length, due to the external pressure p_e with end load, can be used to define the equivalent Poisson's ratio $\overline{\nu_e}$ (cf. eq. [5.7] chap. 5). Putting $A = A_e - A_i$ gives

$$\varepsilon_{a\,pe} = -\frac{F}{E_a A}\left(1 - 2\overline{\nu_e}\right) \tag{E.10}$$

Equation (E.10) has two components—the *axial strain* due to the *axial force* (F),

$$\varepsilon_{aa} = -\frac{F}{E_a A} \tag{E.11}$$

and the *axial strain* due to the *radial force* ($2FL/r_e$) acting on the tube outer surface (radius r_e) over length L,

$$\varepsilon_{ar} = 2\overline{\nu_e}\frac{F}{E_a A} \tag{E.12}$$

Hence, the *axial elongation* of length L, owing to the *radial force* ($2FL/r_e$) is given by

$$\varepsilon_{ar} L = 2\overline{\nu_e}\frac{FL}{E_a A} \tag{E.13}$$

Therefore, from the reciprocal theorem, an *axial force* ($2FL/r_e$) induces a *radial displacement* of the tube outer surface:

$$2\overline{\nu_e}\frac{FL}{E_a A} \qquad (E.14)$$

Hence (dividing by $2L/r_e$), an *axial force F* induces a *radial displacement* u_e of the tube outer surface given by

$$u_e = \overline{\nu_e}\frac{Fr_e}{E_a A} \qquad (E.15)$$

inducing a *circumferential strain,*

$$\varepsilon_{ce} = \frac{u_e}{r_e} = \overline{\nu_e}\frac{F}{E_a A} \qquad (E.16)$$

Since an *axial force F* induces an *axial strain* $\varepsilon_a = -F/(E_a A)$ (see eq. [E.11]). Therefore, from equation (E.16):

$$\overline{\nu_e} = -\frac{\varepsilon_{ce}}{\varepsilon_a} \qquad (E.17)$$

where ε_a and ε_{ce} are respectively the axial strain and the external circumferential strain induced by a *pure axial load.*

The preceding argument was developed by considering the effect of an axial compressive load, but the ratio $\varepsilon_{ce} / \varepsilon_a$ can more be more easily obtained from an *axial tension test*.

Appendix F: Curvature of a Tensioned Beam Subject to Generalized Load

Figure F–1 shows a beam under constant tension subject to lateral load f_x and end moments M_0 and M_L. The differential equation governing the deflected shape is

$$EI\frac{d^4y}{dx^4} - T\frac{d^2y}{dx^2} - f_x = 0 \qquad (F.1)$$

Dividing through by T and rearranging leads to

$$-\left(1 - \frac{EI}{T}\frac{d^2}{dx^2}\right)\frac{d^2y}{dx^2} = \frac{f_x}{T} \qquad (F.2)$$

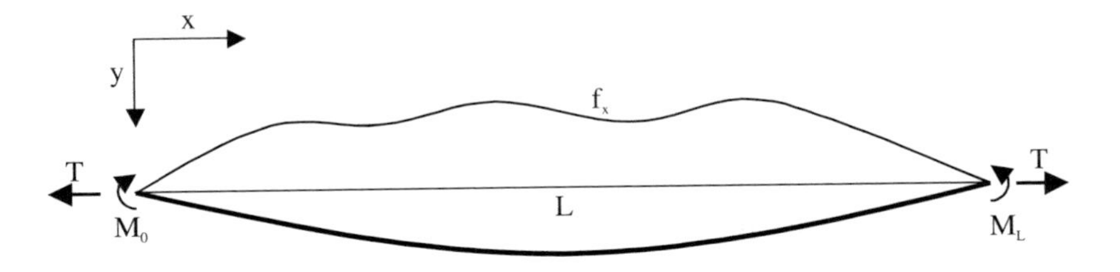

Fig. F–1. *Tensioned beam with lateral load and end moments*

Putting $k = \sqrt{T/EI}$, using the operator D for d/dx (and D^2 for d^2/d^2x, etc.), and noting that the curvature $(1/R)_x = -d^2y/dx^2$ allows equation (F.2) to be rewritten as follows:

$$\left(1 - \frac{D^2}{k^2}\right)\left(\frac{1}{R}\right)_x = \frac{f_x}{T} \qquad (F.3)$$

Putting

$$H_x = \left(1 + \frac{D^2}{k^2} + \frac{D^4}{k^4} + \cdots\right)\frac{f_x}{T} \tag{F.4}$$

allows the solution to equation (F.3) to be written as

$$\left(\frac{1}{R}\right)_x = B_1 \frac{\sinh kx}{\sinh kL} + B_2 \frac{\sinh k(L-x)}{\sinh kL} + H_x \tag{F.5}$$

where B_1 and B_2 are constants of integration that can be found from the solution at $x = 0$ and $x = L$. Noting that 1/R = M/EI, at $x = 0$, equation (F.5) leads to equation (F.6) and, hence, the value of B_2 given by equation (F.7):

$$\left(\frac{1}{R}\right)_0 = \frac{M_0}{EI} = B_2 + H_0 \tag{F.6}$$

$$B_2 = \frac{M_0}{EI} - H_0 \tag{F.7}$$

Further, at $x = L$, equation (F.5) leads to equation (F.8) and, hence, the value of B_1 given by equation (F.9):

$$\left(\frac{1}{R}\right)_L = \frac{M_L}{EI} = B_1 + H_L \tag{F.8}$$

$$B_1 = \frac{M_L}{EI} - H_L \tag{F.9}$$

Substitution for B_1 and B_2 in equation (F.5) gives

$$\left(\frac{1}{R}\right)_x = \left(\frac{M_0}{EI} - H_0\right)\frac{\sinh k(L-x)}{\sinh kL} + H_x + \left(\frac{M_L}{EI} - H_L\right)\frac{\sinh kx}{\sinh kL} \tag{F.10}$$

where H_x is given by equation (F.4). H_0 and H_L are the values of H_x at $x = 0$ and $x = L$.

Application to Parabolic Load

For the parabolic load given by equation (F.11), function H_x, as defined by equation (F.4), is given by equation (F.12):

$$f_x = \left[\sqrt{f_0} + \left(\sqrt{f_L} - \sqrt{f_0}\right)\frac{x}{L}\right]^2 \qquad (F.11)$$

$$H_x = \frac{f_x}{T} + \frac{2}{T}\left(\frac{\sqrt{f_L} - \sqrt{f_0}}{kL}\right)^2 \qquad (F.12)$$

If functions G_0 and G_L are respectively defined as

$$G_0 = \frac{M_0}{EI} - H_0 = \frac{M_0}{EI} - \frac{f_0}{T} - \frac{2}{T}\left(\frac{\sqrt{f_L} - \sqrt{f_0}}{kL}\right)^2 \qquad (F.13)$$

$$G_L = \frac{M_L}{EI} - H_L = \frac{M_L}{EI} - \frac{f_L}{T} - \frac{2}{T}\left(\frac{\sqrt{f_L} - \sqrt{f_0}}{kL}\right)^2 \qquad (F.14)$$

then

$$\left(\frac{1}{R}\right)_x = G_0 \frac{\sinh k(L-x)}{\sinh kL} + \left[\frac{f_x}{T} + \frac{2}{T}\left(\frac{\sqrt{f_L} - \sqrt{f_0}}{kL}\right)^2\right] + G_L \frac{\sinh kx}{\sinh kL} \qquad (F.15)$$

Appendix G: Riser Bundle Pipe Moments between Guides

Figure G–1 shows an individual pipe of a riser bundle, between two guides. The pipe has bending stiffness EI. It is subject to a lateral load per unit length f and an axial effective tension T, which may be positive or negative.

The spans between guides are of equal length L_g and are short enough for the lateral load 'f' and the axial force T to be considered constant over several spans. The problem is to find an expression for the distribution of moments *in the pipe* as a function of position between the guides. M_g is the moment in the pipe at the guides.

Symmetry implies that pipe slope (dy/dx) is equal to bundle slope at the guides. Hence, at $x = 0$, $dy/dx = L_g/2R_{bundle}$, and at $x = L_g$, $dy/dx = -L_g/2R_{bundle}$, where $1/R_{bundle}$ is the bundle curvature. That allows the constants of integration to be found.

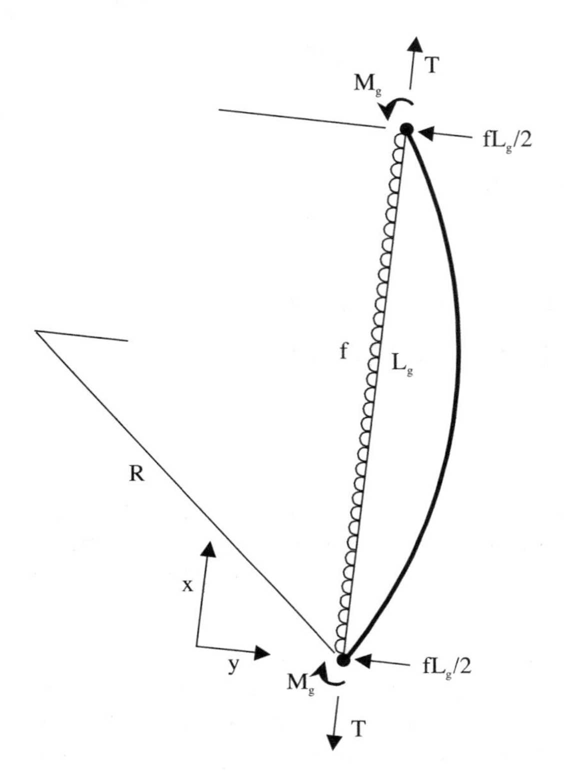

Fig. G–1. *Forces on an individual pipe between guides*

Pipe under Tension (T)

The basic moment equation of the individual pipe in figure G–1 is given by

$$-EI\frac{d^2y}{dx^2} = M_x = M_g + \frac{fL_g}{2}x - \frac{f}{2}x^2 - Ty \qquad (G.1)$$

where all values (EI, M_x, M_g, f, and T) refer to the individual pipe. Putting $k = \sqrt{T/EI}$ and dividing by T allows equation (G.1) to be rearranged as follows:

$$\left(1 - \frac{1}{k^2}\frac{d^2}{dx^2}\right)y = \frac{1}{T}\left(M_g + \frac{fL_g}{2}x - \frac{f}{2}x^2\right) \qquad (G.2)$$

Differentiation of equation (G.2) gives

$$\left(1-\frac{1}{k^2}\frac{d^2}{dx^2}\right)\frac{dy}{dx}=\frac{f}{T}\left(\frac{L_g}{2}-x\right) \tag{G.3}$$

The solution to equation (G.3) is symmetrical about $x = L_g/2$ and can be written in the form of equation (G.4), which yields equation (G.5) by differentiation, where B is a constant to be found:

$$\frac{dy}{dx}=B\sinh\left[k\left(\frac{L_g}{2}-x\right)\right]+\frac{f}{T}\left(\frac{L_g}{2}-x\right) \tag{G.4}$$

$$\frac{d^2y}{dx^2}=-Bk\cosh\left[k\left(\frac{L_g}{2}-x\right)\right]-\frac{f}{T} \tag{G.5}$$

At $x = 0$, the slope $dy/dx = L_g/2R_{\text{bundle}}$. Hence, from equation (G.4),

$$B=\left[\left(\frac{1}{R_{\text{bundle}}}-\frac{f}{T}\right)\right]\frac{L_g}{2\sinh\left(kL_g/2\right)} \tag{G.6}$$

Thus from equation (G.5):

$$\frac{d^2y}{dx^2}=-\left(\frac{1}{R_{\text{bundle}}}-\frac{f}{T}\right)\frac{kL_g}{2}\left\{\frac{\cosh\left[k\left(\frac{L_g}{2}-x\right)\right]}{\sinh\left(\frac{kL_g}{2}\right)}\right\}-\frac{f}{T} \tag{G.7}$$

Since $M_x/EI = -d^2y/dx^2$, equation (G.7) leads to

$$\frac{M_x}{EI}=\left(\frac{1}{R_{\text{bundle}}}-\frac{f}{T}\right)\frac{kL_g}{2}\left\{\frac{\cosh\left[k\left(\frac{L_g}{2}-x\right)\right]}{\sinh\left(\frac{kL_g}{2}\right)}\right\}+\frac{f}{T} \tag{G.8}$$

Putting $M_{pbc} = EI/R_{bundle}$, where M_{pbc} is the pipe moment due to bundle curvature $1/R_{bundle}$ and using equation (G.8) leads to

$$\frac{M_x}{M_{pbc}} = \left[1 - \left(\frac{f}{T}\right) R_{bundle}\right] \frac{kL_g}{2} \left\{ \frac{\cosh\left[k\left(\frac{L_g}{2} - x\right)\right]}{\sinh\left(\frac{kL_g}{2}\right)} \right\} + \frac{f}{T} R_{bundle} \quad (G.9)$$

Pipe under Compression (*F*)

For the pipe under compression the equations are very similar to the above. The basic moment equation is the same as equation (G.1), with $F = -T$:

$$-EI\frac{d^2y}{dx^2} = M_x = M_g + \frac{fL_g}{2}x - \frac{fx^2}{2} + Fy \quad (G.10)$$

where all values (EI, M_x, M_g, f, and F) refer to the individual pipe. Putting $k = \sqrt{F/EI}$ and dividing by F allows equation (G.10) to be rearranged as follows:

$$\left(1 + \frac{1}{k^2}\frac{d^2}{dx^2}\right)y = -\frac{1}{F}\left(M_g + \frac{fL_g}{2}x - \frac{f}{2}x^2\right) \quad (G.11)$$

Using similar reasoning to that presented following equation (G.2) leads to the following solution:

$$\frac{d^2y}{dx^2} = -\left(\frac{1}{R_{bundle}} + \frac{f}{F}\right)\frac{kL_g}{2}\left\{ \frac{\cos\left[k\left(\frac{L_g}{2} - x\right)\right]}{\sin\left(\frac{kL_g}{2}\right)} \right\} + \frac{f}{F} \quad (G.12)$$

Since $M_x/EI = -d^2y/dx^2$, equation (G.12) leads to

$$\frac{M_x}{EI} = \left(\frac{1}{R_{\text{bundle}}} + \frac{f}{F}\right)\frac{kL_g}{2}\left[\frac{\cos\left(\frac{kL_g}{2} - kx\right)}{\sin\left(\frac{kL_g}{2}\right)}\right] - \frac{f}{F} \qquad (G.13)$$

Putting $M_{pbc} = EI/R_{\text{bundle}}$, where M_{pbc} is the pipe moment due to bundle curvature $1/R_{\text{bundle}}$, and using equation (G.13) leads to

$$\frac{M_x}{M_{pbc}} = \left(1 + \frac{f}{F}R_{\text{bundle}}\right)\frac{kL_g}{2}\left\{\frac{\cos\left[k\left(\frac{L_g}{2} - x\right)\right]}{\sin\left(\frac{kL_g}{2}\right)}\right\} - \frac{f}{F}R_{\text{bundle}} \qquad (G.14)$$

Pipe under Zero Axial Load ($T = F = 0$)

For the case of zero axial force, the basic equation is as follows:

$$-EI\frac{d^2y}{dx^2} = M_x = M_g + \frac{fL_g}{2}x - \frac{fx^2}{2} \qquad (G.15)$$

where all values (EI, M_x, M_g, and f) refer to the individual pipe. Integration of equation (G.15) gives the slope:

$$\frac{dy}{dx} = -\frac{1}{EI}\left(M_g x + \frac{fL_g}{4}x^2 - \frac{f}{6}x^3\right) + C \qquad (G.16)$$

At $x = 0$, $dy/dx = L_g/2R_{\text{bundle}}$. Hence the constant C is given by:

$$C = \frac{L_g}{2R_{\text{bundle}}} \qquad (G.17)$$

At $x = L_g$, $dy/dx = -L_g/2R_{bundle}$. Hence,

$$-\frac{L_g}{2R_{bundle}} = -\frac{1}{EI}\left(M_g L_g + \frac{fL_g^{\ 3}}{4} - \frac{fL_g^{\ 3}}{6}\right) + \frac{L_g}{2R_{bundle}} \qquad (G.18)$$

giving

$$M_g = \frac{EI}{R_{bundle}} - \frac{fL_g^{\ 2}}{12} \qquad (G.19)$$

Substituting for M_g in equation (G.19) gives

$$M_x = \frac{EI}{R_{bundle}} - \frac{fL_g^{\ 2}}{12} + \frac{fL_g}{2}x - \frac{fx^2}{2} \qquad (G.20)$$

Putting $M_{pbc} = EI/R_{bundle}$, where M_{pbc} is the pipe moment due to bundle curvature $1/R_{bundle}$, and using equation (G.20) leads to

$$\frac{M_x}{M_{pbc}} = 1 - R_{bundle}\frac{fL_g^{\ 2}}{12(EI)}\left[1 - 6\frac{x}{L_g}\left(1 - \frac{x}{L_g}\right)\right] \qquad (G.21)$$

Appendix H: Catenary Equations

In this appendix, the following are derived:

- Standard equations for inextensible cable catenaries with zero bending stiffness
- The axial stretch of a catenary and its vertical and horizontal projections, resulting from non-zero axial stiffness
- The axial stretch of the flow line on the seabed, ignoring friction
- A numerical formulation of the standard large-angle differential equation for catenaries with bending stiffness

Cable Catenary Equations

Figure H–1 shows the axes and the forces acting on a simple catenary, with origin at the touchdown point (TDP). Angle θ is measured from the horizontal; T is the catenary effective tension, with vertical component V and horizontal component H; w is the apparent weight per unit length; and s is the arc length measured from the TDP.

The vertical force at any point is equal to the apparent weight of the suspended length ($V = ws$), measured from the TDP. Also, at any point, $ds/dx = T/H$, and $dy/dx = V/H$. The following relationships can be deduced from the top-end forces in figure H–1:

$$\frac{ds}{dx} = \frac{T}{H} = \frac{\sqrt{H^2 + V^2}}{H} = \frac{\sqrt{H^2 + (ws)^2}}{H} = \sqrt{1 + \left(\frac{ws}{H}\right)^2} \qquad (H.1)$$

$$\frac{dy}{dx} = \frac{V}{H} = \frac{ws}{H} = \tan\theta \tag{H.2}$$

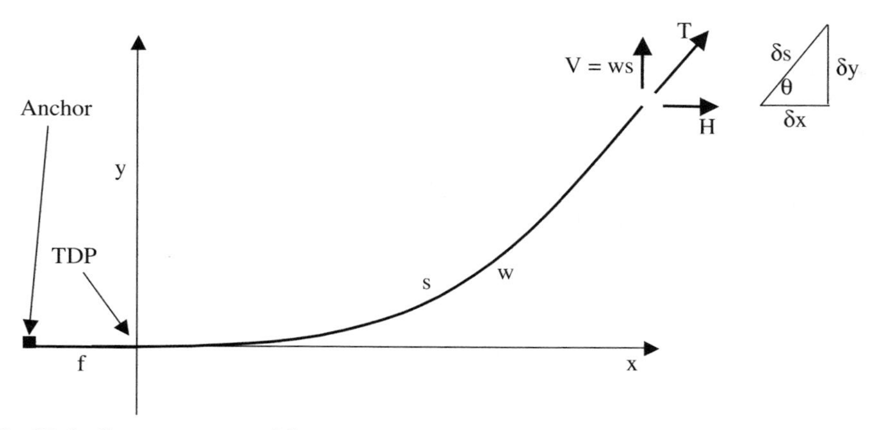

Fig. H–1. *Catenary axes and forces*

The standard catenary equations can be derived from equations (H.1) and (H.2). Equation (H.1) leads to equation (H.3), which when evaluated leads to equation (H.4):

$$\int_0^s \frac{ds}{\sqrt{1+(ws/H)^2}} = \int_0^x dx \tag{H.3}$$

$$\frac{ws}{H} = \sinh\left(\frac{wx}{H}\right) \tag{H.4}$$

Substitution of equation (H.4) into equation (H.1) gives equation (H.5):

$$\frac{T}{H} = \cosh\left(\frac{wx}{H}\right) \tag{H.5}$$

Equations (H.2) and (H.4) lead to equation (H.6) and, hence, to equation (H.7):

$$\int_0^y dy = \int_0^x \sinh\left(\frac{wx}{H}\right) dx \tag{H.6}$$

$$\frac{wy}{H} + 1 = \cosh\left(\frac{wx}{H}\right) \tag{H.7}$$

Therefore, from equations (H.5) and (H.7),

$$\frac{T}{H}=\frac{wy}{H}+1=\cosh\left(\frac{wx}{H}\right) \tag{H.8}$$

The curvature ($1/R = d\theta/ds$) is given by differentiating the last equality of equation (H.2), which gives equations (H.9) and (H.10):

$$\frac{w}{H}\frac{ds}{d\theta}=\sec^2\theta=\frac{T^2}{H^2} \tag{H.9}$$

$$\frac{1}{R}=\frac{d\theta}{ds}=\frac{wH}{T^2} \tag{H.10}$$

Curvature is greatest at the origin (i.e., the TDP), where $T = H$ and, hence, $1/R = w/H$.

Extensible Catenary Equations

For an extensible cable catenary, the axial stretch *on deployment* will modify the values of x, y, s from their unstretched values x_u, y_u, s_u. The axial stretch Δs is obtained by integrating the axial strain ε_a over the suspended length, as given by

$$\Delta s=\int_0^s \varepsilon_a\, ds \tag{H.11}$$

The changes in the horizontal and vertical projections resulting from the stretch are denoted Δx and Δy, respectively, and are given by equations (H.12) and (H.13):

$$\Delta x=\int_0^s \varepsilon_a \cos\theta\, ds \tag{H.12}$$

$$\Delta y=\int_0^s \varepsilon_a \sin\theta\, ds \tag{H.13}$$

From equation (5.7) of chapter 5, the change in axial strain $\Delta\varepsilon_a$ can be expressed by equation (H.14) for anisotropic pipes, with thermal strains added:

$$\Delta\varepsilon_a = \frac{1}{EA}\left[\Delta T + \left(1-2\overline{\nu_i}\right)\Delta p_i A_i - \left(1-2\overline{\nu_e}\right)\Delta p_e A_e\right] + \alpha\Delta t \quad (H.14)$$

where $\overline{\nu_i}$ and $\overline{\nu_e}$ are the internal and external equivalent Poisson's ratios (for isotropic pipes, $\overline{\nu_i} = \overline{\nu_e} = \nu$). In equation (H.14), ΔT, Δp_i and Δp_e, and Δt are the respective changes in effective tension, internal and external pressure, and temperature at any particular point, between the two sets of conditions; α is the coefficient of thermal expansion; A_i and A_e are the internal and external cross-sectional areas; and EA is the axial stiffness.

If the pipe is assumed *initially* to be horizontal, at ambient temperature, subject to zero tension (T = 0) and pressures (p_i = p_e = 0), then from equation (H.11), the axial strain, ε_a, at any point, can be expressed in terms of the *final* values of the tension, pressure, and temperature (T, p_i, and p_e, t) at that point, where temperature t is measured with respect to the initial ambient temperature:

$$\varepsilon_a = \frac{1}{EA}\left[T + \left(1-2\overline{\nu_i}\right)p_i A_i - \left(1-2\overline{\nu_e}\right)p_e A_e\right] + \alpha t \quad (H.15)$$

The internal and external pressures will vary with depth and are given by equations (H.16) and (H.17):

$$p_i = p_{ti} + \rho_i g(d-y) \quad (H.16)$$

$$p_e = \rho_e g(d-y) \quad (H.17)$$

where ρ_i and ρ_e are the internal and external fluid (mass) densities, p_{ti} is the top-end internal pressure, d is the total depth, y is the height above the seabed, and g is the gravitational constant.

A parameter B, independent of depth, is defined as follows:

$$B = \frac{\left(1-2\overline{\nu_i}\right)\rho_i g A_i - \left(1-2\overline{\nu_e}\right)\rho_e g A_e}{w} \quad (H.18)$$

where w is the apparent weight per unit length. By use of this definition, equation (H.15) can then be rewritten as

$$\varepsilon_{\mathrm{a}} = \frac{1}{EA}\left[T + \left(1 - 2\overline{\nu_{\mathrm{i}}}\right)p_{\mathrm{ti}}A_{\mathrm{i}} + Bw(d - y)\right] + \alpha t \qquad (H.19)$$

From equation (H.8), $w(d - y) = (T_{\mathrm{t}} - T)$, where T_{t} is the top tension; thus, equation (H.19) becomes equation (H.20), which can be rearranged as equation (H.21):

$$\varepsilon_{\mathrm{a}} = \frac{1}{EA}\left[T + \left(1 - 2\overline{\nu_{\mathrm{i}}}\right)p_{\mathrm{ti}}A_{\mathrm{i}} + B(T_{\mathrm{t}} - T)\right] + \alpha t \qquad (H.20)$$

$$\varepsilon_{\mathrm{a}} = \frac{1 - B}{EA}T + \frac{T_{\mathrm{t}}B + \left(1 - 2\overline{\nu_{\mathrm{i}}}\right)p_{\mathrm{ti}}A_{\mathrm{i}}}{EA} + \alpha t \qquad (H.21)$$

If a constant value of temperature t is assumed, then in equation (H.18), the effective tension T is the only parameter to vary along the length of the catenary.

Change in the horizontal projection Δx, resulting from pipe stretch

Note from figure H–1, that $T \cos\theta = H$ and $\delta s \cos\theta = \delta x$; thus, equations (H.12) and (H.21) give equation (H.22):

$$\Delta x = \int_0^s \frac{1 - B}{EA}H\,ds + \int_0^x \left[\frac{T_{\mathrm{t}}B + \left(1 - 2\overline{\nu_{\mathrm{i}}}\right)p_{\mathrm{ti}}A_{\mathrm{i}}}{EA} + \alpha t\right]dx \qquad (H.22)$$

Hence,

$$\Delta x = \left(\frac{1 - B}{EA}H\right)s_{\mathrm{u}} + \left[\frac{T_{\mathrm{t}}B + \left(1 - 2\overline{\nu_{\mathrm{i}}}\right)p_{\mathrm{ti}}A_{\mathrm{i}}}{EA} + \alpha t\right]x_{\mathrm{u}} \qquad (H.23)$$

Change in the vertical projection Δy, resulting from pipe stretch

Note from figure H–1 that $T \sin\theta = ws$ and $\delta s \sin\theta = \delta y$; thus, equations (H.13) and (H.21) give equation (H.24):

$$\Delta y = \int_0^s \frac{1-B}{EA} ws\, ds + \int_0^y \left[\frac{T_t B + (1-2\bar{\nu}_i) p_{ti} A_i}{EA} + \alpha t \right] dy \qquad (H.24)$$

Hence,

$$\Delta y = \left(\frac{1-B}{2EA} w \right) s_u^2 + \left[\frac{T_t B + (1-2\bar{\nu}_i) p_{ti} A_i}{EA} + \alpha t \right] y_u \qquad (H.25)$$

Total axial stretch Δs

Equations (H.11) and (H.21) give

$$\Delta s = \int_0^s \frac{1-B}{EA} T\, ds + \int_0^s \left[\frac{T_t B + (1-2\bar{\nu}_i) p_{ti} A_i}{EA} + \alpha t \right] ds \qquad (H.26)$$

By putting $u = wx/H$, then $T = H \cosh u$, and $ws/H = \sinh u$ (see eqq. [H.4] and [H.5]. Hence $ds/du = (H/w) \cosh u$. Substitution of these expressions in equation (H.26) leads to equation (H.27), which, when integrated, yields equation (H.28):

$$\Delta s = \frac{1-B}{EA} \frac{H^2}{w} \int_0^u \cosh^2 u\, du + \left[\frac{T_t B + (1-2\bar{\nu}_i) p_{ti} A_i}{EA} + \alpha t \right] s_u \qquad (H.27)$$

$$\Delta s = \frac{1-B}{EA} \frac{H^2}{w} \left(\frac{\cosh u \sinh u + u}{2} \right)_0^u + \left[\frac{T_t B + (1-2\bar{\nu}_i) p_{ti} A_i}{EA} + \alpha t \right] s_u \qquad (H.28)$$

By substitution for cosh u, sinh u and u in equation (H.28) leads to:

$$\Delta s = \frac{1-B}{2EA} \frac{H^2}{w} \left(\frac{T_t}{H} \frac{ws_u}{H} + \frac{wx_u}{H} \right) + \left[\frac{T_t B + (1-2\bar{\nu}_i) p_{ti} A_i}{EA} + \alpha t \right] s_u \qquad (H.29)$$

This simplifies to:

$$\Delta s = \frac{1}{2EA}\left[(1+B)T_t s_u + (1-B)Hx_u\right] + \left[\frac{(1-2\overline{\nu}_i)p_{ti}A_i}{EA} + \alpha t\right]s_u \qquad (H.30)$$

Flow-line stretch Δf

The flow-line stretch is given by

$$\Delta f = \int_0^{f_u} \varepsilon_a \, df \qquad (H.31)$$

If friction with the soil is neglected, the flow-line *strain* is given by equation (H.21), with the effective tension $T = H$. Hence, the strain equation is as follows:

$$\Delta f = \int_0^{f_u} \frac{1-B}{EA} H \, df + \int_0^{s_u} \left[\frac{T_t B + (1-2\overline{\nu}_i)p_{ti}A_i}{EA} + \alpha t\right] df \qquad (H.32)$$

Hence, the flow-line stretch Δf is given by

$$\Delta f = \left[\frac{H(1-B) + T_t B + (1-2\overline{\nu}_i)p_{ti}A_i}{EA} + \alpha t\right] f_u \qquad (H.33)$$

where f_u is the unstretched flow-line length.

Numerical Analyses

The basic large-angle deflection equation derived in appendix A (see eq. [A.11]) is repeated here as equation (H.34):

$$EI\frac{d^3\theta}{ds^3} - T\frac{d\theta}{ds} + w\cos\theta + f(s) = 0 \qquad (H.34)$$

Equation (H.34) can be represented numerically by different algorithms in terms of the coordinates of adjacent nodes. For the case of equally spaced nodes, it can be expressed at node *n,* in terms of the angles at five nodes: the node itself, the two nodes immediately above it, and the two nodes immediately below it, as shown in figure 12–4.

The angle θ with the horizontal is known at each node, and e is the element length. Numerical expressions for the terms in equation (H.34) can be derived as follows, where subscripts "$n + 2$," "$n + 1$," "n," "$n - 1$," and "$n - 2$" are used for the values at the five nodes.

The curvature $(d\theta/ds)_n$ at node n is equal to the mean rate of change of the angle between nodes $n - 1$ and $n + 1$. Hence,

$$\left(\frac{d\theta}{ds}\right)_n = \frac{\theta_{n+1} - \theta_{n-1}}{2e} \quad (H.35)$$

By use of subscript "$n + 1/2$" to represent the midpoint between nodes n and $n + 1$ and subscript "$n - 1/2$" for the midpoint between nodes $n - 1$ and n, the curvature $(d\theta/ds)$ at those midpoints is given by equations (H.36) and (H.37):

$$\left(\frac{d\theta}{ds}\right)_{n+1/2} = \frac{\theta_{n+1} - \theta_n}{e} \quad (H.36)$$

$$\left(\frac{d\theta}{ds}\right)_{n-1/2} = \frac{\theta_n - \theta_{n-1}}{e} \quad (H.37)$$

The second differential $(d^2\theta/ds^2)_n$ at node n is the mean rate of change of $d\theta/ds$; hence,

$$\left(\frac{d^2\theta}{ds^2}\right)_n = \frac{(d\theta/ds)_{n+1/2} - (d\theta/ds)_{n-1/2}}{e} = \frac{\theta_{n+1} - 2\theta_n + \theta_{n-1}}{e^2} \quad (H.38)$$

Likewise, the third differential $(d^3\theta/ds^3)_n$ is equal to the mean rate of change of $d^2\theta/ds^2$, as given by equation (H.39), which leads to equations (H.40) and (H.41):

$$\left(\frac{d^3\theta}{ds^3}\right)_n = \frac{\left(d^2\theta/ds^2\right)_{n+1} - \left(d^2\theta/ds^2\right)_{n-1}}{2e} \quad (H.39)$$

$$\left(\frac{d^3\theta}{ds^3}\right)_n = \frac{1}{2e}\left(\frac{\theta_{n+2} - 2\theta_{n+1} + \theta_n}{e^2} - \frac{\theta_n - 2\theta_{n-1} + \theta_{n-2}}{e^2}\right) \quad (H.40)$$

$$\left(\frac{d^3\theta}{ds^3}\right)_n = \frac{\theta_{n+2} - 2\theta_{n+1} + 2\theta_{n-1} - \theta_{n-2}}{2e^3} \tag{H.41}$$

Therefore, from equations (H.35) and (H.41), equation (H.34) can be expressed numerically at node n by

$$EI\left(\frac{\theta_{n+2} - 2\theta_{n+1} + 2\theta_{n-1} - \theta_{n-2}}{2e^3}\right) - T_n\left(\frac{\theta_{n+1} - \theta_{n-1}}{2e}\right) + w\cos\theta_n + f(s)_n = 0 \tag{H.42}$$

Appendix I: Damped Axial Vibrations

Damped-vibration equations are very long to derive. Only the basic equations and the solutions are given in this appendix.

Riser with Distributed Damping

This case corresponds to figure 14–1*d* of chapter 14. The damped-wave equation (eq. [14.20]) is repeated here as equation (I.1):

$$m\frac{\partial^2 u}{\partial t^2} = mc^2\frac{\partial^2 u}{\partial x^2} - \lambda_D\frac{\partial u}{\partial t} \qquad (I.1)$$

The top-end displacement ($u_0 \sin \omega t$) is satisfied by the *imaginary* part of equation (I.2):

$$u_{x,t} = u_0\left[B_1 e^{-jKx} + (1 - B_1)e^{jKx}\right]e^{j\omega t} \qquad (I.2)$$

where B_1 and K are complex numbers ($K = K_1 - jK_2$). Substitution of equation (I.2) into (I.1) yields the values of K_1 and K_2:

$$K_1 L = \left[(\omega L / c)\cos\phi\right] / \sqrt{(\cos 2\phi)},$$

and

$$K_2 L = \left[(\omega L / c)\sin\phi\right] / \sqrt{(\cos 2\phi)},$$

where

$$\tan 2\phi = (\lambda_D L / mc)/(\omega L / c).$$

Consideration of the equilibrium of the stab at $x = L$ leads to equation (I.3), the solution of which is given by equation (I.4):

$$M\left(\frac{\partial^2 u}{\partial t^2}\right)_L + mc^2\left(\frac{\partial u}{\partial x}\right)_L + \lambda_B\left(\frac{\partial u}{\partial t}\right)_L = 0 \qquad (I.3)$$

$$u_{x,t} = u_0 \left\{ \begin{array}{l} \left[\begin{array}{l} \left(e^{2K_2x} + \alpha \sinh K_2 x\right)\cos K_1 x \\ + \beta\left(\cosh K_2 x\right)\sin K_1 x \end{array} \right] \sin \omega t \; + \\ \left[\begin{array}{l} \left(e^{2K_2x} + \alpha \cosh K_2 x\right)\sin K_1 x \\ - \beta\left(\sinh K_2 x\right)\cos K_1 x \end{array} \right] \cos \omega t \end{array} \right\} \qquad (I.4)$$

where

$$\alpha = \frac{F_1}{F_3}$$

$$\beta = \frac{F_2}{F_3}$$

$$F_1 = G - (P + 1 - N)e^{2K_2L}$$

$$F_2 = Q\cos(2K_1L) + N\sin(2K_1L)$$

$$F_3 = (1 - N)\cosh(2K_2L) + P\sinh(2K_2L) - G$$

$$G = Q\sin(2K_1L) - N\cos(2K_1L)$$

$$N = \frac{1 - D_1^2 - D_2^2}{2}$$

$$P = D_1\cos\phi - D_2\sin\phi$$

$$Q = D_1\sin\phi + D_2\cos\phi$$

$$D_1 = \frac{\lambda_B}{mc}\sqrt{\cos(2\phi)}$$

$$D_2 = \frac{M\omega}{mc}\sqrt{\cos(2\phi)}$$

Since $T_x = mc^2\,\partial u/\partial x$, differentiation of equation (I.4) yields the dynamic-tension function:

$$T_{x,t} = mc^2 u_0 \left\{ \begin{array}{l} \left[\begin{array}{l} (K_2 e^{K_2 x} + \overline{\alpha K_2 + \beta K_1} \operatorname{Cosh} K_2 x)\cos K_1 x \\ +(-K_1 e^{K_2 x} - \overline{\alpha K_1 - \beta K_2} \operatorname{Sinh} K_2 x)\sin K_1 x \end{array} \right] \sin \omega t \; + \\ \left[\begin{array}{l} (K_1 e^{K_2 x} + \overline{\alpha K_1 - \beta K_2} \operatorname{Cosh} K_2 x)\cos K_1 x \\ +\,(K_2 e^{K_2 x} + \overline{\alpha K_2 + \beta K_1} \operatorname{Sinh} K_2 x)\,\sin K_1 x \end{array} \right] \cos \omega t \end{array} \right\}$$

(I.5)

At x, the *amplitudes* U_x of the displacement and T_x of the dynamic tension are then given by

$$U_x = U_0 \sqrt{R_x - S_x} \qquad (I.6)$$

$$T_x = mc\omega U_0 \sqrt{\frac{R_x + S_x}{\cos(2\phi)}} \qquad (I.7)$$

where

$$R_x = \overline{1+\alpha}\, e^{2K_2 x} + \left(\frac{\alpha^2 + \beta^2}{2} \right) \cosh(2K_2 x) \qquad (I.8)$$

$$S_x = \left(\frac{\alpha^2 + \beta^2}{2} + \alpha \right) \cos(2K_1 x) - \beta \sin(2K_1 x) \qquad (I.9)$$

Response with Damping at Equivalent End ($x = L'$)

The wave equation (eq. [13.2]) and the top-end imposed displacement (at $x = 0$) are satisfied by equation (I.10), which yields equations (I.11) and (I.12):

$$u_{x,t} = U_0\left[\left(B_2 \sin\frac{\omega x}{c} + \cos\frac{\omega x}{c}\right)\sin\omega t + B_3 \sin\frac{\omega x}{c}\cos\omega t\right] \quad (I.10)$$

If a damping coefficient (λ') is applied at the lower equivalent end (at $x = L'$), then consideration of equilibrium at $x = L'$ leads to equation (I.11), which allows the constants B_2 and B_3 to be found:

$$\lambda'\left(\frac{\partial u}{\partial t}\right)_{L'} + mc^2\left(\frac{\partial u}{\partial x}\right)_{L'} = 0 \quad (I.11)$$

The solution to equation (I.11) is given by

$$u_{x,t} = U_0\left\{\left[\frac{\cos\frac{\omega L'}{c}\cos\frac{\omega(L'-x)}{c} + \gamma^2\sin\frac{\omega L'}{c}\sin\frac{\omega(L'-x)}{c}}{\cos^2\frac{\omega L'}{c} + \gamma^2\sin^2\frac{\omega L'}{c}}\right]\sin\omega t + \left(\frac{-\gamma\sin\frac{\omega x}{c}}{\cos^2\frac{\omega L'}{c} + \gamma^2\sin^2\frac{\omega L'}{c}}\right)\cos\omega t\right\} \quad (I.12)$$

Differentiation then gives the dynamic tension, since $T_{x,t} = mc^2(du/dx)_{x,t}$:

$$T_{x,t} = mc\omega U_0\left\{\left[\frac{\cos\frac{\omega L'}{c}\sin\frac{\omega(L'-x)}{c} - \gamma^2\sin\frac{\omega L'}{c}\cos\frac{\omega(L'-x)}{c}}{\cos^2\frac{\omega L'}{c} + \gamma^2\sin^2\frac{\omega L'}{c}}\right]\sin\omega t + \left(\frac{-\gamma\cos\frac{\omega x}{c}}{\cos^2\frac{\omega L'}{c} + \gamma^2\sin^2\frac{\omega L'}{c}}\right)\cos\omega t\right\} \quad (I.13)$$

At x, the *amplitudes* U_x of the displacement and T_x of the dynamic tension simplify to

$$U_x = U_0 \sqrt{\frac{\cos^2 \frac{\omega(L'-x)}{c} + \gamma^2 \sin^2 \frac{\omega(L'-x)}{c}}{\cos^2 \frac{\omega L'}{c} + \gamma^2 \sin^2 \frac{\omega L'}{c}}} \tag{I.14}$$

$$T_x = mc\omega U_0 \sqrt{\frac{\sin^2 \frac{\omega(L'-x)}{c} + \gamma^2 \cos^2 \frac{\omega(L'-x)}{c}}{\cos^2 \frac{\omega L'}{c} + \gamma^2 \sin^2 \frac{\omega L'}{c}}} \tag{I.15}$$

Equivalent Damping: Energy Dissipated per Cycle

The object is to find the value of the damping factor γ of the equivalent system, shown in figure 14–1*f*, that dissipates the same quantity of energy per cycle as the system with distributed damping defined by the coefficients λ_D and λ_B, shown in figure 14–1*e*. The energy dissipated per cycle at the equivalent lower end ($x = L'$) of the equivalent system owing to λ' is given by

$$E_{\text{cycle},\lambda'} = \lambda' \int_0^{2\pi/\omega} U_{L'}^2 \sin^2 \omega t \, dt = \lambda' U_{L'}^2 \left(\frac{\pi}{\omega}\right) \tag{I.16}$$

Similarly, the energy dissipated per cycle at the lower extremity ($x = L$) of the riser with distributed damping owing to λ_B is given by

$$E_{\text{cycle},\lambda_B} = \lambda_B U_L^{\ 2} \left(\frac{\pi}{\omega}\right) \tag{I.17}$$

The energy dissipated per cycle ($\delta E_{\text{cycle},\lambda_D}$) over length δx at point x owing to λ_D is given by

$$\delta E_{\text{cycle},\lambda_D} = \lambda_D U_x^{\ 2} \delta x \int_0^{2\pi/\omega} \sin^2(\omega t + \phi)\, \delta t = \lambda_D U_x^{\ 2} \delta x \left(\frac{\pi}{\omega}\right) \tag{I.18}$$

where U_x is the amplitude of the displacement at x. Hence, the energy dissipated over the total length L per cycle ($E_{\mathrm{cycle},\lambda_\mathrm{D}}$) is given by

$$E_{\mathrm{cycle},\lambda_\mathrm{D}} = \lambda_\mathrm{D} \int_0^L U_x^2 \, \delta x \left(\frac{\pi}{\omega} \right) \tag{I.19}$$

Equating the energy dissipated by the two systems leads to the requirement

$$\lambda' U_{L'}^2 = \lambda_\mathrm{D} \int_0^L U_x^2 \, \delta x + \lambda_\mathrm{B} U_L^2 \tag{I.20}$$

By substitution from equation (I.14) for $U_{L'}$ and from equation (I.6) for U_x and U_L, the required value of γ (where $\gamma = \lambda'/mc$) is given by

$$\frac{\gamma}{\cos^2(\omega L'/c) + \gamma^2 \sin^2(\omega L'/c)} = \frac{\lambda_\mathrm{D}}{mc} \int_0^L (R_x - S_x) \, \delta x + \frac{\lambda_B}{mc}(R_L - S_L) \tag{I.21}$$

At resonant frequencies, $\cos(\omega L'/c) = 0$, and $\sin(\omega L'/c) = 1$. Hence, equation (I.21) reduces to

$$\frac{1}{\gamma} = \frac{\lambda_\mathrm{D}}{mc} \int_0^L (R_x - S_x) \, \delta x + \frac{\lambda_\mathrm{B}}{mc}(R_L - S_L) \tag{I.22}$$

Appendix J: Notes on Excel Files

List of Files

Excel files are provided on the accompanying CD-ROM so that the reader may explore and verify the findings of many of the chapters, using different data. Ideally, the reader should visualize the files on a computer screen, while progressing through the book. The format of the data and results, with their color codes, has been standardized to facilitate understanding. Some of the files, however, contain denser, tabulated information that is difficult to appreciate fully on a computer screen. Therefore, printable copies of those files are also provided.

Each file as listed here is preceded by the number of the chapter in which the file is mentioned and its purpose is explained. Where several files are referred to in the same chapter, the lowercase letter gives the order in which they are mentioned. The following 17 files are included in the folder *Screen-Files*:

1. *SCR-Example.xls*

2. *Riser-Tensions.xls*

4. *Riser-Stresses.xls*

6. *Tensioned-Beam.xls*

8*a*. *Uniform-Riser.xls*

8*b*. *Segmented-Riser.xls*

9. *SJ-Design.xls*

10. *Bundle-Moments.xls*

11*a*. *TLP-Risers.xls*

11*b*. *Floater-Risers.xls*

12*a*. *TDP-Shift.xls*
12*b*. *Catenary-500.xls*
13. *Fixed-Axial-Vibrations.xls*
14*a*. *Hungoff-Free-Vibrations.xls*
14*b*. *Hungoff-Damped-Vibrations.xls*
15*a*. *Uniform-Transverse-Vibrations.xls*
15*b*. *Segmented-Transverse-Vibrations.xls*

The folder *Print-Files* contains the following four files, which are adapted for ease of printing. Apart from a slightly modified layout and the absence of color, they are identical to the screen versions listed previously. They also allow the user to specify a title.

1. *SCR-Example (print).xls*
2. *Riser-Tensions (print).xls*
4. *Riser-Stresses (print).xls*
12*a*. *TDP-Shift (print).xls*

All files require input data, except for *Catenary-500.xls,* which is included so that the reader may *observe* details of a particular SCR simulation.

File Formats and Color Codes

All the screen versions of the files have been prepared with similar formats and color codes. Required data are very compact and self-explanatory. Only a minimum number of data values are required. Results are generally presented in both tabular and graphical form. In the interest of clarity, results are also kept to the minimum required in order to illustrate specific points raised in the text.

The following colors codes are used:

- *Blue* for titles
- *Brick red* for data requests
- *White* for data values
- *Beige* for principal results, or data values imposed by the file
- *Mauve* for secondary results or information

In many of the files, results of *simplified analysis* are compared with *exact results*. Generally, but not always, the former are *analytical* and the latter are *numerical* results. In the graphs and tables, simplified results are generally given in red, and exact results are given in blue. In the case of *Uniform-Transverse-Vibrations.xls,* five calculation methods are compared; thus, five colors are used.

Numerical Calculations

The Excel files of chapters 8, 9, and 15 include numerical results. They are obtained by applying the finite difference method, in which the riser or stress joint is modeled by 600 elements. Figure J-1 shows a segment of uniform riser represented by elements of equal length *e.*

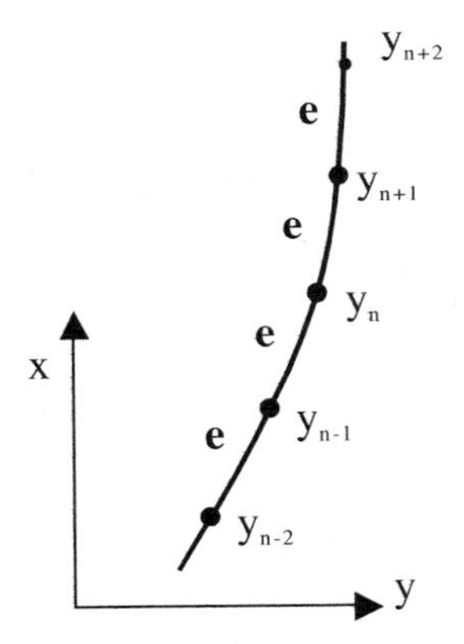

Fig. J–1. *Segment of uniform riser showing five equally spaced nodes*

The numerical method uses equations similar to those given in appendix H, but adapted to small-angle deflections with the vertical. The basic equation to be represented numerically is

$$\frac{d^2M}{dx^2}+\frac{d}{dx}\left(T\frac{dy}{dx}\right)+f(x)=0 \qquad (J.1)$$

At node *n,* the first and second terms of equation (J.1) are given in terms of the lateral offsets of the node itself and the nodes above and below it by equations (J.2) and (J.3), where $T_{n+1/2}$ and $T_{n-1/2}$ are the effective tensions at the midpoints of the elements immediately above and below node *n*:

$$\left(\frac{d^2M}{dx^2}\right)_n = EI\frac{-y_{n+2}+4y_{n+1}-6y_n+4y_{n-1}-y_{n-2}}{e^2} \qquad (J.2)$$

$$\left[\frac{d}{dx}\left(T\frac{dy}{dx}\right)\right]_n = \frac{T_{n+1/2}\left(y_{n+1}-y_n\right)-T_{n-1/2}\left(y_n-y_{n-1}\right)}{e^2} \qquad (J.3)$$

The third term of equation (J.1) is data for the applications of chapter 8 and zero for the stress joints of chapter 9. For the transverse vibrations of chapter 15, $f(x) = -my_n\omega^2$.

Equations (J.2) and (J.3) apply to uniform segments of riser. At the junctions between riser segments with different characteristics, similar equations can be formulated, but they are more complicated because of the discontinuities in the characteristics.

For segmented risers, formulation of the second and third terms of equation (J.1) is still straightforward. By use of the symbols (EI_a, w_a, a) for the characteristics of the bending stiffness, apparent weight, and element length above the junction, and (EI_b, w_b, b) for those below it, the formulation of the second differential of the moment at the junction $(d^2M/dx^2)_j$—in terms of the moments at the junction (M_j), at the node above the junction (M_{j+a}), and at the node below the junction (M_{j-b})—is also straightforward. It is given by equation (J.4), which simplifies to equation (J.5):

$$\left(\frac{d^2M}{dx^2}\right)_j = \frac{\left[\left(M_{j+a}-M_j\right)/a\right]-\left[\left(M_j-M_{j-b}\right)/b\right]}{(a+b)/2} \qquad (J.4)$$

$$\left(\frac{d^2M}{dx^2}\right)_j = \frac{bM_{j+a}-(a+b)M_j+aM_{j-b}}{ab(a+b)/2} \qquad (J.5)$$

The only difficulty is to express the moment M_j at the junction in terms of the lateral offsets y_j, y_{j+a}, y_{j-b} at the junction itself, at the node above the junction, and at the node below the junction. It can be found from the relationship between the angles, the moments, and the curvature. At the junction, the slope ϕ_j and the moment M_j are continuous, whereas the curvature changes abruptly from $(1/R)_a = -d\phi/ds = M_j/EI_a$ to $(1/R)_b = -d\phi/ds = M_j/EI_b$. The moment at the junction (M_j) is finally given by

$$M_j = \frac{-by_{j+a}+(a+b)y_j-ay_{j-b}}{(ab/2)\left\{\left[a/EI_a\right]+\left[b/EI_b\right]\right\}} \qquad (J.6)$$

Index

A

B

C

D

E

F

G

H

I

J

K

L

M

N

O

P

R

S

T

U

V

W

Y